LOLA
GT
THE DNA OF THE FORD GT40
John Starkey

Other motorsport books from Veloce –

First published in March 2022 by Veloce Publishing Limited, Veloce House, Parkway Farm Business Park, Middle Farm Way, Poundbury, Dorchester DT1 3AR, England.
Tel +44 (0)1305 260068 / Fax 01305 250479 / e-mail info@veloce.co.uk / web www.veloce.co.uk or www.velocebooks.com.
ISBN: 978-1-787117-83-9; UPC: 6-36847-01783-5.

British Library Cataloguing in Publication Data – A catalogue record for this book is available from the British Library. Typesetting, design and page make-up all by Veloce Publishing Ltd on Apple Mac. Printed in India by Parksons Graphics.

THE DNA OF THE FORD GT40

John Starkey

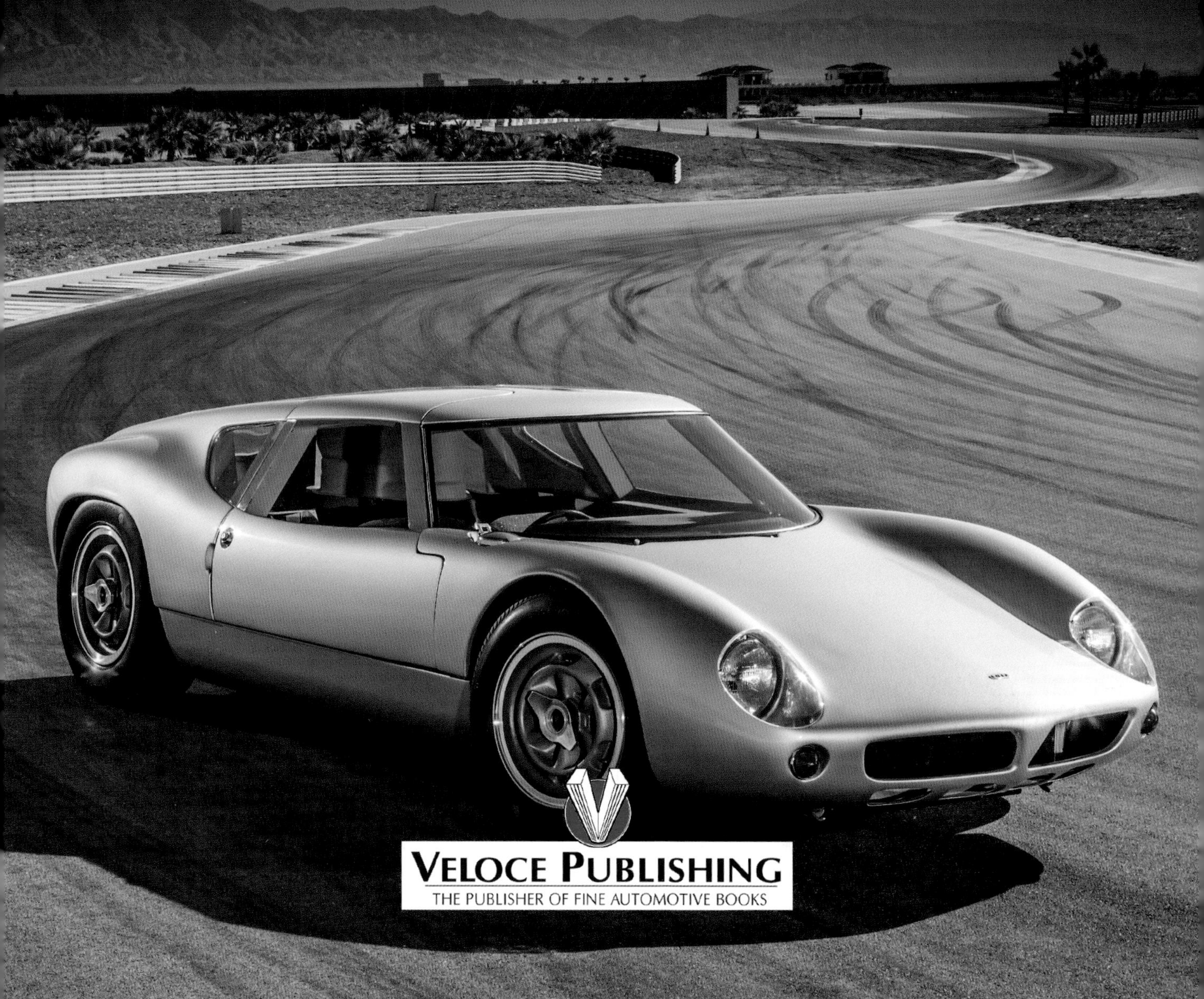

Contents

Acknowledgements

First of all, a big thank you to Allen Grant, who has owned 1963 Lola MK VI GT, chassis number LGT/P, since May in 1965. I also have to thank my long suffering Publisher, Rod Grainger of Veloce Publishing in Dorset, England, for allowing me to twist his arm about publishing this book. Then there is Franco Varani, who hails from Scotland and is the most assiduous person in finding old cars' histories.

Many years ago I used to speak frequently with Laurie Bray, who had been the foreman at Lola Cars Ltd, both at Bromley and at Slough. He gave me invaluable details of just what happened during the times that Lola and Ford co-operated. Thank you Laurie.

Finally, to my wife, Su, who has to put up with a lot of my concentrating on the writing project in hand, (meaning that I don't pay attention to what she is telling me sometimes), plus my rambling on about obscure cars and personalities. Thank you, Su.

Introduction

The 2019 film *Ford versus Ferrari*, which received great public acclaim, detailed (with some Hollywood variations ...) the story of Ford and Ferrari at Le Mans in 1966. There is a scene, early on in the film in which the new Ford GT40, completely formed, is unveiled to the Press in 1964. Little is said of just where this spectacular new car came from, and indeed, the impression is given that this car had been built by the Ford factory in America, which I am sure is what the producers were intending. Nothing could be further from the truth.

This, then, is the *real* story of what became the Ford GT40, the Lola Cars Ltd-built Mk 6 GT coupé of 1963. It was this car that caused Ford of America to enter into a contract with Lola in order to produce the GT40, a very similar car indeed.

Just three Mk 6 GT coupés were built, and they all survive. I am indebted to the owner of one of them, Allen Grant, who has owned and cherished his car (LGT/P the prototype) for nearly 60 years, after buying it from Lola Cars Ltd when he worked for Ford Advanced Vehicles in England at that time. Allen supplied me with much material that I had not previously seen, including a booklet by Roy Lunn of the Ford Motor Company in Dearborn. Lunn was the man who, in 1963, was charged by Ford of America with using Lola's services to design and develop the Ford GT40, for this is what the Lola Mk 6 GT presaged. It is a fascinating story, mainly about one man, Eric Broadley, the head and at that time chief designer and tester of the cars built by Lola.

Eric and his little company, which never produced more than 3000 racecars in total, are a shining example of the sort of British designers and entrepreneurs who set the racing car world on its way to the huge success it has since become.

Chapter 1

Sports car racing in the 1950s

To understand the reasons behind the design and development of the Lola Mk 6 GT car of 1962/3, which would later be developed into the Ford GT40, we have to go back to the start of the 1950s.

Les 24 Heures du Mans, as the French write it, started in 1923. The race was not run between 1940 and 1948 – the Second World War years and immediately afterwards. When it re-started in 1949 it was won by a Ferrari 166MM, for Ferrari's first of a total of nine eventual victories. 'Le Mans'

The C-Type Jaguar won at Le Mans in 1951 and 1953. (Author's collection)

immediately went back to being *the* premier race for sports cars in the world, a title it had gained during the interwar period.

William Lyons, head of the Jaguar Car Company of Coventry, England, decided that it was well worth putting time, effort and money into a bid to win Le Mans, particularly as his new XK120 sports car had featured so well there in 1950, leading at one point, with the best one finishing as high as 12th overall. Lyons firmly believed in the publicity, and subsequent sales success a Le Mans win would bring.

In a determined effort to win at Le Mans, his team designed and developed the 'C' (for Competition) Jaguar, using many components from the XK120. It duly won the race in 1951 and repeated the trick in 1953. Jaguar probably would have won as well in 1952 had it not believed the improbable story that the Mercedes 300SLs, which had been encountered in May at the Mille Miglia, had a higher top speed. In a hurry, without testing properly, Jaguar gave the C-Type a longer nose and tail, alterations that involved changing the header tank of the cooling system. The engines boiled and put them out of the running very early on.

Using its brutal 4.9-litre Tipo 375 plus, Ferrari came back to win again in 1954, but only just, as Jaguar's new D-Type pressed it very hard. Jaguar won again in 1955, the year of the terrible Mercedes 300SLR crash that killed over 80 spectators, and then the Jaguar factory, feeling that it had learned all that Le Mans could teach in terms of car design, racing and marketing, retired.

It was now left to privately run Jaguar D-Types, in the hands of Scottish team Ecurie Ecosse, to win again in 1956 and 1957.

Ferrari returned in style in 1958, and, apart from Aston Martin triumphing in 1959 (a certain Mr Carroll Shelby, who will figure largely in this book, was the co-driver, with Roy Salvadori of the winning DBR1/300), it dominated the race again from 1960 until 1964.

In 1958, the Cooper Cars Company of Surbiton had produced what appeared to be a revolution where racing sports cars was concerned: the Cooper T57 Monaco, a two-seat mid-engined race car. Although it had a Coventry Climax FPF four-cylinder engine of only 1.5-2.0 litres, it was used to great effect by not only the works, but also a slew of customers who

A 1953 Ferrari 375MM, of the type that competed in the Le Mans 24 Hours. (Author's collection)

The D-Type Jaguar won the 24 Hours of Le Mans in 1955, '56 and '57. (Author's collection)

One of the Ferrari factory-entered cars in later life. (Courtesy John O'Steen)

lined up to buy this giant-killing sports car. Lotus was late getting into this market, but by 1960 it was producing its own version of what a mid-engined sports-racing car should be: The Lotus 19.

By 1962, Ferrari was using a mid-engined car, the six-cylindered Dino 206 and the V8-engined 248. By 1963, Ferrari would be using the 250P, another mid-engined sports car. It was obvious which way the wind was blowing ... a revolution in sports car design was happening.

With world attention focused on Le Mans, it was easy to see how important it had become as a marketing tool, something of which the Ford Motor Company in America took note of when starting its new marketing programme, which emphasized performance above all else.

With the coming of the mid-engine revolution in racing cars in the early to mid 1960s, sports car racing was changed for all time. Gone were the glorious sculpted, front-engined Ferraris, Maseratis and Jaguars, and in their place came a new breed of mainly British-designed mid-engined cars, such as the Ford GT40, and, shortly thereafter, the Lotus 30 and 40 and the Lola T70.

1959 Cooper T57 Monaco, chassis number 002, a 'customer' race car. The true forerunner of the mid-engined sports car. (Author's collection)

Chapter 2

Lola Cars Ltd

Although it depends on continuous development, motor racing is arguably a conservative occupation. Generally speaking, designs tend to be refined and developed by their designers, rather than a revolution taking place with each new design.

Hence, up to the late 1950s, with the exception of the Auto Union Grand Prix cars of the mid to late 1930s, successful racing cars tended to be front-engined. The Cooper Car Company of England changed this with the design of its mid-engined cars, both F1 and Sports, during the 1950s, culminating in

Lola in the beginning: Brands Hatch 1957 and the 1172 special waits in the paddock. (Courtesy Lola Archives)

Eric Broadley at work. (Courtesy Lola Archives)

the Formula One cars that Jack Brabham used to win the World Championship in 1959 and 1960.

Most manufacturers were loath to follow the Cooper example, but gradually, on seeing its success, the more perceptive constructors did. One of these was a quantity surveyor named Eric Broadley, from Bromley in Kent, who had founded Lola Cars Ltd, a little company that made and sold racing cars.

Eric and his cousin Graham, who had been an aircraft engine mechanic in the Fleet Air Arm, had built their first racing sports car in 1956, an Austin 7 special (the 'Broadley Special'), powered by a Ford

A Mk 1 being loaded onto its trailer outside the old Lola factory in Bromley, Kent. Eric Broadley can be seen third from the left, and the old workshop's roof can be seen on the left of this picture, with the new workshop on the right. (Courtesy Frank Lugg)

750cc engine, while Eric still worked at his 'normal' job by day. Eric entered his first race at Silverstone and came last ... but it was a start, and encouraged him to set up his own racing car company.

Graduating to the 1172cc class, Eric and Graham built another special later in 1956, and with it, after learning how to tune the Ford engine, Eric began to win races in 1957. Graham's father wouldn't allow Graham to race on a Sunday, so by default Eric became the equipe's driver.

Somehow and somewhere along the way the car acquired the name that was to stick: Lola. Maybe it came from the song – *Whatever Lola wants, Lola gets* – but Eric and Graham Broadley were now devoting less of their time to this special. Fired by ambition, the cousins sold the 1172 special for £600, along with Eric's motorbike, in a quest to swell their funds.

The cousins saved money and started building another, more advanced car that would run in the up-to-1172cc class using the lightweight 1100cc Coventry climax FWA engine. This engine had an overhead camshaft and gave 83bhp, and the car was fitted with independent front and rear suspension. At this time, in early 1958, the up-to-1172cc class

The entrance to the first Lola Cars Ltd workshop, circa 1959. (Courtesy Frank Lugg)

was seen as the first rung on the ladder for young drivers who wished to climb to the top in motor racing, much as Formula Ford would later become.

Eric had looked at the de Dion rear suspension of the all-conquering Lotus 11 and drew a suspension that used the driveshaft as the transverse link, with the top wishbone mounting making up a trailing arm. The differential casing design also featured a light crown wheel and pinion casing, and hub carriers that were cast in aluminium to save weight. At the front was a fabricated double wishbone setup, utilising uprights from the then contemporary Morris Minor saloon car. TR2 drum brakes were fitted inboard at the rear, with finned Alfin drums, and this completed the running gear. An Austin A30 gearbox with Lotus close-ratio gears connected the FWA engine to a BMC 4.55:1 differential, and T43 Cooper 15-inch wheels were used.

All of this was mounted in a strong tube-frame chassis of bronze-welded 20 gauge steel square tubing. The cousins built the car at the local garage of Rob Rushbrook, a friend who had been to watch them race and came away impressed. Rob had a machine shop that Eric had asked to use at night to save money.

The Coventry Climax FWA engine cost £250, and Eric then visited Morris Gomm's body shop, Arch Motors, and quickly discovered a discarded front nose that was being built for a Tojeiro chassis, but which had not been collected. This front-end bodywork was modified to fit the new chassis, and the rest of the body followed in July 1958: the car was finished and registered (600 DKJ) for the street. It was unpainted and had taken just six months to complete.

The cousins then took their pride and joy to the Crystal Palace race track on 5th July 1958, where Eric drove it in the London Trophy race, using the two heats as a test session. Two weeks after this, Eric placed second at Snetterton, beaten only by a Lotus 11 driven by Keith Greene. The Lola then went to Brands Hatch, where it caused a mini sensation. In the first heat Eric won outright with a margin over the second place finisher of 24 seconds, but in the final he was black flagged for erratic driving. Eric promptly entered the race for sports cars up to 1500cc and finished fourth.

Three weeks later at the next meeting at Goodwood, Eric hit a bank whilst avoiding a back marker who spun in front of him. With black eyes and swollen knees he went in to work the next day, only to receive an ultimatum from his boss: the job or motor racing. Eric chose motor racing.

With the proceeds of the sale of the 1100cc sports car to Peter Gammon, and with £1000 capital put up by his father, Eric started Lola Cars Ltd by moving into Maurice Gomm's premises in Byfleet in Surrey, where three Mk 1 cars were built and sold. Eric had been commuting to Byfleet from Chelsfield, near Orpington, and now he called Rob Rushbrook to see whether he could use his premises at Bromley in Kent, as the commuting was becoming too much for him. Rob owned some land next to his garage and, drawing on Eric's contacts in the building business, new premises were soon completed. At this time, Lola had several orders to build Mk 1s, owing to its success on the track. In 1960, a total of 19 Mk 1s were built and sold from the new premises in Bromley.

The Lola Mk 1 really established Lola Cars as a manufacturer to watch. 35 examples of this car were produced over the next four years.

All of these were produced in the little workshop situated in Bromley, Kent, behind a row of terraced houses. About once every four houses was an archway to allow access and, situated over the archway nearest to the Lola 'factory,' was a hand-painted sign: 'Lola Cars Ltd.'

Frank Lugg, who worked at Lola Cars Ltd in the 'early days' of 1959 remembered:

"Eric Broadley had just the right qualities and temperament to do what he did. Apart from his design ability he was a meticulous craftsman doing a job quickly to a high standard, but never fussing unnecessarily.

"The old Amos and Rushbrook workshop, where I first worked, was a more or less square corrugated iron shed, which had been a garage since 1904. It was bitterly cold in the workshop; seemingly even colder inside than outside. Sometimes our hands would stick to the chassis tubes in the early mornings of winter. Rob Rushbrook employed one or two garage mechanics who often had to work

A Lola Mk 1 used in Historic racing today. (Author's collection)

outside in dreadful conditions during this period before the new workshop was built. We worked at anything that was required; in my case mainly making up chassis components which were then nickel-bronze welded together in the very accurate jigs which Eric made.

"Working for Eric Broadley was really very civilized: he and Rob Rushbrook were even-tempered with a sense of humour (rather sharp in Eric's case!). They always asked rather than gave orders, and nobody swore, not even the mechanics.

"We were not paid a lot and had to work extra time, Eric sending me up to Birmingham and Coventry in his Ford van after lunch one day, which meant that I didn't return home until eight o'clock that evening. I was hungry and anxious to get back and Eric said that his van was never the same again!"

The little company went on to build more race cars: the Lola Marks 2 to 4, built from 1959 to 1962, were single-seat cars, ranging from front-engined Formula Junior to mid-engined Formula One cars with Coventry Climax V8 engines of 1.5 litres fitted.

Chapter 3

The Lola GT Mk 6 coupé

In the latter part of 1962, Eric Broadley designed the Lola GT Mk 6 to new FIA rules that had been issued that year. He later said:

"We built the Lola Mk 6 coupé because it seemed like a good thing to do at the time. By late 1962 a big American engine in the back of a small GT car was, to me, the logical way to go, and time has proven that we definitely made the right choice. We had to work like hell to get the prototype ready for the Racing Car Show in January 1963, and the car wasn't much more than a shell when we finally got it to the show and put it on the display stand."

Eric was intending to build both open and closed versions, and exhibited the GT (closed) version at the Racing Car show in London in January 1963, where it attracted many admiring glances and was the undisputed hit of the show.

To get the new Lola to the show, Eric was said to

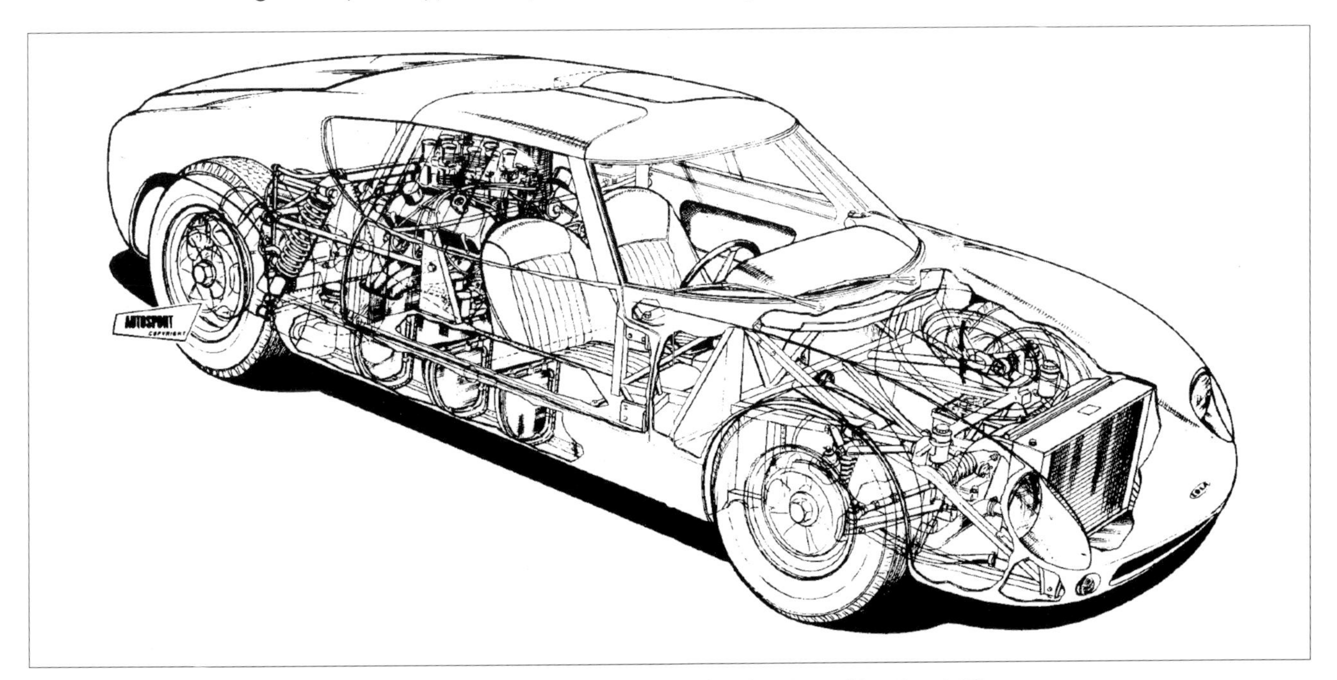

The Lola GT Mk 6 cutaway drawing by Theo Page. (Courtesy LAT)

The first chassis being delivered to Lola Cars Ltd. (Courtesy Lola Archives)

The prototype GT Mk 6 being built up at the factory. (Courtesy Lola Archives)

have spent 52 hours with no sleep, and had suffered a power failure at his workshop. This resulted in the new Mk 6 arriving two days after the show opened, and it was far from complete, having no engine installed, and no driveshafts either, and with a ground clearance that was two inches lower than intended in race trim.

"It makes the Ferrari Berlinetta look like a Lorry" said Nick Syrett, secretary of the British Racing and Sports Car Club, as the Lola was pushed into Olympia, and most of the visitors to the show seemed to endorse this view ... (*Car and Driver*, July 1963).

The Motoring Press was ecstatic. *Motor Sport* said:

"Eric Broadley has built a fantastic Lola coupé with a 4.7-litre Ford V8 engine behind the seats and a Colotti transmission behind the rear axle. The first of these incredible machines, in mock-up form, was seen at the BRSCC Racing Car Show, and two of these cars have been entered for Le Mans, which should cause the French scrutineers to choke on their Gauloises. Previously Maserati held the crown for building 'wild rides,' but if this Lola works, then the crown must surely pass to Eric Broadley."

Autosport's correspondent wrote:

A view of the engine bay of the prototype Lola GT Mk 6. (Courtesy Lola Archives)

"The extremely low-slung, attractive car should give a good account of itself in Prototype races this season. Powered by a 4.2, or 4.7-litre Ford V8 engine, the little coupé has a top speed of 180mph, which makes it much faster than any of the Grand Prix cars competing this season."

Motor Racing featured a two-page spread on the GT Mk 6 and concluded:

The seldom seen nose section of the GT Mk 6. Here can be seen the radiator, battery, steering rack and master cylinders. (Courtesy Ford Archives)

"Surely the most compact GT car of its class ever built ... No final price has been fixed, but it will probably be something over £5000 for the competition GT."

Car and Driver, after discussing how the FIA had instituted rulings over the years to make sports and GT cars slower, discussed the new rules for GT prototypes that had been instituted for the Le Mans 24 Hours and were starting in 1963. Its correspondent wrote:

"However, no capacity limits were imposed. Either this was an oversight or it was assumed the high weight limits of 1923 pounds for a car of 3001 to 5000cc would discourage the production of large-engined cars.

"With all this the FIA no doubt felt that they had seen the last of all those fast exciting cars which have drawn vast crowds to Le Mans and the Nürburgring year after year. But they had reckoned without the ingenuity of people like Eric Broadley, designer and managing director of the little Lola company at Bromley on the south eastern outskirts of London, England.

"When the GT prototype regulations came through Broadley already had a car on the stocks scheduled for completion in January, in time for the London Racing Car Show. Because of all the changes that had to be made, it was a late arrival at Olympia,

Another view of the Ford engine and Colotti gearbox installed in the prototype Lola GT Mk 6 in Lola's workshop. (Courtesy Lola Archives)

but when it finally materialized it was generally acclaimed to be the star of the show and the first of a new breed of motorcar, even though it was incomplete and had never turned a wheel."

The Lola GT Mk 6 coupé prototype featured a monocoque chassis made of duralumin and sheet steel, the floor having two 'D' shaped sills/rocker panels, with an aluminium tank inside to hold

Above & opposite: Four photos taken by Eddie Hull, a Ford of America engineer, at the Lola factory in Bromley, showing details of the GT Mk 6. (Courtesy Ford Archives)

The Lola GT Mk 6, minus its nose, being pushed into position at the Racing Car Show in London in 1963. Eric Broadley is at the right rear, helping to push the car onto its stand. (Courtesy Lola Archives)

the fuel, and a tube subframe at the front that supported the front suspension and radiator. Inside each fuel tank/sill were four cast formers, made of cast magnesium, which took the roof structure and door frame attachment bolts.

The second and third cars built, known as the 'production' cars (LGT/1 and LGT/2), featured more aluminium in the monocoque tub, for lightness, than the prototype. The rocker panel/sills were now sealed inside with a liquid rubber type sealant that was used then on military aircraft, to hold fuel on each side (30 gallons total capacity). The same design features would be repeated for Eric Broadley's next design, the Lola T70. The first T70s featured a mainly steel monocoque, but all the following Marks 2, 3 & 3b monocoques would be mainly built of aluminium, with rubber bag-type tanks as used in military aircraft. The fuel tanks of the T70 Mk 1 were very similar in construction to those of the production GT Mk 6s. The front

A model stands proudly by the prototype Lola GT Mk 6 at the Racing Car Show in 1963. (Courtesy Lola Archives)

With taillights from the road-going Ford Cortina fitting in as if they were designed for it, the shapely rear of the GT Mk 6 is shown here at the Racing Car Show in 1963. (Courtesy Lola Archives)

The Lola GT Mk 6 at the London Racing Car show, January 1963. (Courtesy Lola Archives)

At last! The GT Mk 6, although late, on its stand at the Racing Car Show, London Olympia, in January 1963. (Courtesy Lola Archives)

John Wyer (left), and Eric Broadley stand beside the Lola GT Mk 6 after testing in late 1963. (Courtesy Lola Archives)

subframe was slightly altered for the production Mk 6, to aid ease of manufacturing.

A tubular frame to support the roof and provide roll-over protection was mounted above the monocoque centre section. It was hidden by a fibreglass skin above it and another one below. Where driver comfort was concerned, the Lola GT Mk 6 coupé had a very spacious cockpit, capable of accommodating drivers of up to six feet five inches in height.

At the rear, the engine was supported by extensions to the side panniers, and the rear suspension and Colotti gearbox were hung off the rear crossmember/transverse bridge structure. Suspension was by upper and lower wishbones and coil spring damper units at the front, and a top link plus reversed lower wishbones and radius arms at the rear. The uprights were cast in magnesium. The outboard mounted B Type Girling 11-inch diameter disc brakes, with aluminium calipers, were covered by 15-inch diameter magnesium wheels mounted on centre-lock hubs. These were 6.5 inches wide by 15 inches diameter at the front, and eight inches wide by 15 inches diameter at the rear. Steering was by rack and pinion from a Saab, with just two-and-a-half turns taking the front wheels from lock-to-lock, and the glass reinforced plastic (GRP) bodywork was designed by John Frayling (production designer of the Lotus Elite's bodywork), and made by Peter Jackson's Specialised Mouldings, based in Upper Norwood.

Both front and rear bodywork sections were removable for easy access to the suspension, spare wheel and steering at the front, and to the suspension, engine and gearbox at the rear. The roof was cut away on each side, with the doors, although hinged at the front, providing the ease of entry that normally went with a gullwing arrangement (the GT40 has similar style doors). Between the roof cutaways was a neat little NACA duct that fed air into the carburettors.

In the engine bay nestled the Carroll Shelby-tuned Ford 260in^3 (4262cc) engine, which received its fuel via twin Bendix fuel pumps to its four downdraught Weber carburettors. The engine had been sent to Lola by Jacques Passino of Ford USA, at Eric's request. Accompanying the engine was Shelby's right-hand man, Phil Remington, who assisted with the engine's installation. Phil Remington also helped with the test programme of the Lola GT Mk 6, once Ford had taken it over and, on one memorable test occasion, sat next to Bruce McLaren, a can of fuel on his lap and connected to the engine, whilst McLaren drove, attempting to find out what was the problem with the induction system that the Lola/Ford, as it was known at the time, was suffering from.

The Ford Fairlane engine's 260bhp was transmitted to the rear wheels via an Italian-made Colotti-Francis type 37 four-speed gearbox, with centre-mounted gearshift in the cockpit. This central gear lever selected the gears via Bowden cables, rather than the usual metal rod, and this was to prove troublesome during the GT Mk 6's career. The clutch was a Borg and Beck twin-plate diaphragm type. With an expected 350bhp at 7000rpm from the more highly tuned versions, performance was seen as being sensational by the standards of the time.

Continued on page 28

The Lola GT Mk 6 on the street, with Eric Broadley at the wheel. (Courtesy Lola Archives)

This photograph, and those on the following pages, show construction details of the car. (Courtesy Lola Archives)

About the engine, *Sports Car*, in its May 1963 edition reported:

"Broadley's choice of a light-alloy, 4.2-litre Ford V8 engine for his GT car was no doubt inspired by the phenomenal success of the Ford-engined AC Cobra, which surely must be the fastest production sports car in the world ... As fitted to the Lola GT, it had four downdraught Weber carburettors, and power outputs of from 320 to 350bhp, depending on the state of tune, were quoted around the show stand."

Height was just 40 inches, similar to that of the later Ford GT40. Wheelbase was 92 inches, track was 52 inches.

All up weight, according to the Lola specification sheet, was quoted at "about 1800 pounds." The rules specified a minimum of 1923 pounds for a car of between 3001 to 5000cc at Le Mans, so it must be assumed that the quoted weight was without tyres, driver and fuel. Either that, or ballast was needed.

The body design, together with Lola's track record up to then, saw to it that the new GT Mk 6 was declared the star of the show. After the Olympia show, Eric intended that the Lola GT should prove itself on the track as quickly as possible, and the car was readied for a first outing at Goodwood for the Easter Monday race, with Hugh Dibley set to drive it. Time ran out, however, and the work needed to ready the Mk 6 for this date was too much. It had been Eric Broadley's intent to build six each of the open and closed versions, but events overtook this plan.

Almost immediately, the new Lola GT Mk 6 showed its promise by finishing on the same lap as the leaders, at its first race at the Silverstone International meeting on 11th May 1963, despite having started from the back of the grid. *Autosport* wrote:

"Ferrari refused to give John Surtees permission to drive the Lola, and at the last minute Tony Maggs took over, starting from the back of the grid, never having even sat in the car before." And: "Maggs took the Lola around in fine style, picking up nine places in one lap." And finally: "Up and up went the Lola, with Maggs lapping at around 100mph. The car was unfortunately fitted with a single downdraught carburettor to its Ford V8 engine, and not the competition 'four barrels' pattern."

John Surtees had lapped the car in practice at 1.50:2 the previous day, and Tony Maggs, as we have seen, was a last minute substitution. Due to a lack of practice, Maggs started from the back of the grid but still managed to finish ninth, at one point lapping four seconds faster than Surtees had in practice.

In a later interview with Dave Friedman about the Lola GT Mk 6, Tony Maggs recounted:

"Even by today's standards, that Lola was a most beautiful car. When I was asked to drive it at Silverstone, I had never really seen the car before, much less sat in it.

"John Surtees was supposed to drive the car but because John was then under contract to Ferrari and they put their foot down, he had to back out at the last moment. I was called up on the public address system 20 minutes before the start of the race, and Eric and John asked if I could help them out. As a friend of John Surtees, I think that I just happened to be free at the right time to help them out of their difficulty. Luckily, I had my driving suit and helmet with me."

At the Nürburgring in May for the 1000km race later that month, Maggs and Bob Olthoff were forced to retire when the wheelnuts loosened and a wheel fell off, and then, after that had been fixed, and the car had shown great promise, the distributor drive sheared. In between practice and the race, holes were cut at the front of the rear fenders to allow more air to get to the carburettors. After the car finally stopped, Bob Olthoff had to hit a policeman in order to be allowed to climb the spectator fence, so that he could walk back to the pits!

Bob Olthoff said:

"I very much enjoyed driving the Lola GT Mk 6 at the Nürburgring in 1963. That was my first drive in a V8 powered car, and it was the most powerful car that I had driven then. That Lola was a very quick motor car, even though it wasn't sorted out at that point. The car handled really well but we had problems with the gearbox selection and with wheels falling off. The drive gear on the distributor breaking is what finally put us out of the race."

Neil P Albaugh was at the Nürburgring (he took one of the photos shown here), and recalled:

"I saw this car race at the Nürburgring in Germany in 1963. When I was looking over the paddock wall I overheard two British mechanics talking about it.

Tony Maggs driving the prototype Lola GT Mk 6 at Silverstone during its first race in May 1963. (Courtesy Lola Archives)

Silverstone in May 1963, and Tony Maggs and Eric Broadley discuss Maggs' opinion of the Lola after its first race. (Courtesy Lola Archives)

'Have you seen the new Lola? It makes the Ferrari Berlinetta look like a lorry!' It was a beautiful car but did not finish the race. The track announcer informed the crowd that the Lola had stopped near the Pflantzgarten with 'tyre trouble.' In truth, the Colotti transaxle had failed."

The first production car, LGT/1, was intended for the Le Mans 24 Hours, and featured many development changes over the prototype. The fuel holding sponsons on each side were now formed of aluminium, and the engine was fitted with four downdraught Weber IDA 48s on the 289 (4.7-litre) Ford V8. The engine weighed 506 pounds, but when one considers the Jaguar XK engine, which had been heavily used by the sports car racers in the 1950s/early 1960s, that engine had weighed 600 pounds and probably produced not more than a 'real' 240bhp in racing tune. Even in standard tune, the 289 HiPo engine produced 271bhp, and in racing tune it started out at 350bhp ...

The Lola GT Mk 6's suspension had been further developed and strengthened because of what the team had learned at the Nürburgring, and what became the later GT40's suspension was fitted. This meant that the bodywork had to be redesigned by Peter Jackson's Specialised Mouldings to meet FIA regulations concerning windscreen and front bodywork height. In addition, the radiator position was moved, and new cooling ducts and a new exhaust system were fitted. For Le Mans, the car was painted in British Racing Green, with a white longitudinal stripe.

The Colotti-Francis gearbox of the GT Mk 6 was a weak point in the design. It was later established that the real fault was in the Bowden cable gearshift itself, rather than the gearbox, although the Colotti gearbox was swiftly replaced by one from ZF when the Ford GT40 was developed. Certainly, as Laurie Bray, Lola's foreman, later said: "That Bowden cable setup never worked properly." Time had run out for the team, and the gear ratios installed for the Le Mans circuit were too low. Despite this, and the resultant low top speed down the Mulsanne Straight, the lap times were good, due to the car's cornering ability.

At Le Mans, Richard Attwood was to drive the car, together with David Hobbs. Another GT Mk 6 was entered, to be driven by Tony Maggs and John Love, this one having a Chevrolet V8 fitted, but time ran out and the uncompleted car did not take part in the race. It was later on sold to John Mecom for Augie Pabst to drive.

Donald Parker, a race car builder from South Africa, helped Eric with the trip to France. This was Lola's first trip to Le Mans, and Eric later admitted that he had little idea of what was required. Thankfully, Donald spoke French, and a quick call to Le Mans confirmed the basic requirements of a Le Mans race team – food, electric lighting and a marquee to work under. It was decided that Donald would leave for Le Mans ahead of the team, and so, with a newly acquired old Ford Thames van, a marquee, tools, lights and numerous provisions, he set off for the circuit.

Scrutineering came the following day (Tuesday), but the Lola was not there yet, missing its scheduled slot of early afternoon. That night Eric sent a telegram to Donald to advise the problem – Eric, Tony Southgate and the team were still working long into the night to get the car finished – so Donald checked out the lie of the land and struck up a good relationship with the Secretary of the Le Mans race, Raymond Accat. Raymond advised Donald that he wanted the Lola to enter and would make concessions. Donald's good French, and the fact that Raymond knew one of Donald's relatives, helped the situation. Under the Le Mans rules, all cars should have passed scrutineering by 4pm on Wednesday June 12.

That morning, after Tony Southgate had finished hacksawing and then re-fiberglassing the front wheelarches, Eric loaded the car onto the Midland Racing Partnership transporter for the team to take across the English Channel between Ramsgate and Dunkirk. However, fog held up the crossing, and the transporter was severely delayed. To get around the delay, while the transporter had no alternative than to go by sea, Eric decided to take the Lola entry by air ferry. With his heart in his mouth, Eric watched as a ferry employee drove the Lola onto the plane, the employee mastering the Colotti box and parking it with precision. Eric later said "We should have taken him on as a race driver on the spot!"

From Dunkirk, Eric drove the Lola entry a further 280 miles to Le Mans, with team mechanic Don Beresford in the passenger seat, and a toolkit on Don's lap. On the French leg of the journey there were some problems with the gearchange, which

Continued on page 35

The Lola GT Mk 6 at the Nürburgring before it was forced to retire. It had shown itself to be the equal of any of its competitors. (Courtesy Lola Archives)

A fine colour shot of the Lola GT Mk 6 at speed at the Nürburgring in 1963. (Courtesy Neil P Albaugh)

The Lola GT Mk 6 in the pits, being serviced during the Nürburgring 1000km of 1963. (Courtesy Lola Archives)

The Lola on the pit straight at the Nürburgring in May 1963. (Courtesy Lola Archives)

The nimble Lola GT Mk 6 at the Nürburgring, where it was driven by Tony Maggs and Bob Olthoff. (Courtesy Lola Archives)

John Mecom's Lola GT Mk 6, LGT/2 at the hairpin at Sebring in the 12-hour race there in 1964. The Lola was a retirement. (Courtesy Lola Archives)

Just arrived ... the first production Lola GT Mk 6 at Le Mans in June 1963. Hours of work awaited the crew to get it ready for the start. (Courtesy Lola Archives)

Eric Broadley sits in the driver's seat of the Lola GT Mk 6, LGT/1, at Le Mans. This was the first production car, and it looks as if it has just arrived, Eric driving all the way from Slough to Le Mans. (Courtesy Ken Wells)

did not bode well for the race, and Eric and Don had to undertake some running repairs when the suspension they had adjusted for the road came loose, delaying the journey further.

Tony Southgate had joined Lola Cars Ltd as the draughtsman in the drawing office in 1962, starting off drawing the Formula Juniors, after Eric Broadley had sketched out the basic design, and then the suspension geometry. About the GT Mk 6, he later said in an interview in the book *Lola T70 Owners' Workshop Manual*, by Haynes Publishing:

"When we'd finished the GT car it was literally driven out of the garage straight to Le Mans. It had never turned a wheel before that, which was quite a challenge, but you did that kind of thing in those days. Eric drove it there and the mechanic got in the spare seat with a toolbox on his knee and they drove off."

4pm came and still the Lola had not shown up. Donald negotiated numerous extensions every half hour, and early on Thursday morning, Eric and Don finally arrived with LGT/1. Getting the GT to Le Mans was one thing, but the scrutineering, let alone the race, was still to come. So late on Thursday morning, two days after the scheduled time, Eric drove the Lola to the scrutineering bay. The GT Mk 6 had never been through such a process, and this was an anxious time for all. A while later, a dejected and sleep-deprived Eric Broadley returned to the team. According to Don Beresford, Eric said nothing, but ordered a stiff drink, which was unusual for him. While downing the drink, he uttered the words, "The car's failed scrutineering, it's all over." The list of 'failures' was not long, but looked as final as Eric's words sounded. The main problems were lack of rear vision because of the huge central tunnel air intake, inadequate rear bodywork around the rear wheels, and the fuel tanks were too large.

Enter Lola volunteer Donald Parker again, who with the blessing of Raymond Accat and Jacques Finance, the head of the ACO, gained the team an extension of three hours. This was not much time, but the team was now used to completing the impossible. The key to overcoming the problems was Specialised Moulding's Peter Jackson. If Eric had not invited Peter along as part of the team, he would have just packed up and gone home. But Peter was there, and he soon set to work with his expert use of fibreglass mat and resin. The team removed the central air intake tunnel, and Peter and the team set about fabricating a lower collection box in fibreglass and aluminium. Instead of feeding air to the carburettors from the roof, this box would now be fed by air piped from a vent on the left rear of the car, resulting in significantly improved rearward vision. At the same time, the two Dons started fabricating metal rear wheelarch extensions, which were bonded to the bodywork and were almost unnoticeable in relation to the original design.

Finally, the fuel tank problem was overcome by inserting numerous plastic bottles through vents in each tank. As it turned out, all of this work was not quite completed in the allotted three hours but, when the scrutineers visited the Lola marquee, they were so impressed by the progress made in such a short time, and because the organisers were anxious to have a large field of cars, that a little more time was allowed to finish the job. Shortly afterwards, a tired and unshaven Eric Broadley was shaking hands with Jacques Finance, and Eric even found time for a smile.

After these frantic last minute changes, the Lola took the start and was up to 12th by midnight. Sadly, it crashed out of the race after the drivers complained of difficult gear selection in the early morning hours. David Hobbs:

"That Lola was fast! Certainly no-one passed us on the straights. Trouble was, the gear selection was a weak point, and at times I could select two gears at once. In the early morning, I missed a gear and spun, hitting a bank and that was the end of our Le Mans."

The Lola GT Mk 6 went well at Le Mans for over eight hours. (Courtesy Lola Archives)

Richard Attwood, son of the Jaguar distributor for Wolverhampton in England, and who in 1970 won Le Mans in a Porsche 917, together with Hans Hermann, said:

"When I drove the Lola GT Mk 6 at Le Mans in 1963 it was a very revolutionary car. That sort of rear-engine GT car had never been done before in sports car racing, and it was really exciting to be in on the ground floor of something so special. The worst part of the car was the bloody gearbox. The way that the gears were changed in that car gave us trouble from the start. I remember that the gearbox was reliable, but because of the way that you had to change the gears, it was impossible for it to work in that car. That Lola was hugely quick considering the tyres that we had then."

David Hobbs, in his book *Hobbo* said:

"The scrutineers at Le Mans had basically little technical knowledge, and they seemed to take great delight in measuring dimensions with boxes and so when a box wouldn't fit, we would hammer out the area, cut out new holes and so on.

"Despite a lack of running, that Lola was immensely quick, and in the race we got it up to eighth position by around midnight. During the middle of the night, I did the second fastest lap of the race, with only the Phil Hill/Willy Mairesse Ferrari quicker, but the car's difficult Italian Colotti gearbox change was a weak link and continually troublesome. Soon after midnight, the gearbox itself went wrong, and of course you could not replace it as teams do today. You had to fix it. Malcolm Malone, the chief mechanic, was our saviour ... With cars screaming past flat out a few feet away and no pit wall then, Malcolm scrambled around under the car with the gearbox in pieces on bits of rag on the ground. Eventually, after nearly two hours, he got it back together, but now we were down to three gears.

"Unfortunately, the gearbox jammed a couple of hours later, at around 5am as I was going down to the Esses, and the car spun, putting us out of the race ... I went on to race at Le Mans 20 times, and that was my only significant 'off'."

Eric Broadley later said:

"We never had any time to test the car before we ran it at Silverstone in May. We had problems with a wheel falling off at the Nürburgring, and we were rushing about trying to get a second car ready for Le Mans. That didn't happen however, and I drove the one car we had finished to Le Mans with a mechanic riding with me. When we finally got on the track after a terrible row with the French scrutineers, the car went bloody well until the gearbox packed it in and David Hobbs crashed. We were very happy with the overall concept of that car, and we were thrilled by the way that it ran. Looking back, it would have been nice to carry on with that project.

"Our intention was to develop the Mk 6 into a full fledged racing car, but we would've had to raise considerable financing at that time, and I'm not sure how we could've done that. When Ford came into the picture in mid-1963, it was an answer to our financial problems and we went with it. Our choice was a bit of a pity really, because the Ford GT, as a race car, was a bit of a backward step. That car was heavy and it was made of steel, but it became a good project; it solved some of our money problems, and it wasn't too bad to work on."

Tony Southgate later on recounted how the Le Mans entry and subsequent drive in the race impressed the Ford personnel who were there to develop the Lola into what became the Ford GT40. He went on to say:

"As soon as the deal was done, in mid-1964 [1963 ... Author], suddenly all these Ford engineers descended on us. Where we worked in Bromley, there was no room for them. My office was about ten-feet square, and that was it. And the workshop was about 2000 square feet, and everybody almost touched one another. So Ford got a place in Slough that they could put their people in with about four of their senior engineers who were flash blokes, old blokes by my standard. They were twice my age in flashy, shiny suits, and I thought 'Well, I'm not going to get much of a look in here.'"

Tony left and moved to Brabham's factory before returning to Lola to help design and draw the T60 and the T70.

Laurie Bray, who later became the workshop development manager, remembered:

"I started working at Lola in September 1963. Prior to that, I had worked at Aston Martin, on the racing side under John Wyer. I looked after John Ogier's Aston Martin DB4GT Zagato that Jimmy Clark drove, and also the John Coombs' car for Roy Salvadori. Jean Kerguen's car was also works-supported by us, too. I helped with the development and

Eric Broadley (left) and John Wyer look pensive as they stand behind the Lola GT Mk 6 prototype LGT/P, outside the Lola works some time after June 1963. (Courtesy Lola Archives)

maintenance of the 'Project' cars also, the 212, 214 and 215.

"When Aston Martin withdrew from racing, Don Beresford, who worked with Eric Broadley (as well as Rob Rushbrook), told him that I might be available to work for Lola, so Eric invited me to Bromley for a chat. When I went to Bromley (which was halfway across London from where I then lived, in Sunbury on Thames), Eric also asked me if I knew anyone else who might be available, and I mentioned Aston Martin racing mechanics Andy Prescott and Terry Hadley, who also came to Lola to work.

"In those days, Lola's workshop in Bromley was just big enough to hold two cars and two to three machines. There was a single paraffin heater in the middle of the shop, a bucket of water to wash your hands in, and a toilet!"

In 1963, Lola sold the third GT Mk 6 coupé (LGT/2) to John Mecom, who became the Lola distributor in America before Ford concluded its deal with Lola Cars Ltd. As it was not then ready, he had the repaired Le Mans GT Mk 6 taken to Brands Hatch for Augie Pabst to drive in the Guards Trophy, during which the Ford 260 engine blew. Mecom had his car engined with a Traco Chevrolet 327in^3 V8, and painted the Lola in his colours of metallic blue with a white longitudinal stripe.

Continued on page 42

CONFIDENTIAL

Ford Motor Company

FORD DIVISION

Intra-Company Communication

OFFICE OF P. F. LORENZ

1963 SEP 12 PM 2:13

GENERAL OFFICE

September 12, 1963

To: Mr. P. F. Lorenz

cc: Mr. L. A. Iacocca

From: F. E. Zimmerman, Jr.

Subject: Mecom Racing Team

The purpose of this communication is to outline the current status of our discussions with John Mecom, Jr.

As you know, in early July the Division entered into an agreement with Eric Broadley, designer of the Lola Car, to work exclusively with us on the overall development of a Ford G.T. vehicle. Prior to this arrangement with him, Mr. Broadley had accepted an order for a Lola Car from John Mecom and was at the time in the process of completing this vehicle. Since this was the only Lola vehicle ever produced, it was felt that it would be beneficial to us to endeavor to make an arrangement with Mecom that would permit us to re-purchase this car to use in connection with our G.T. development work. Such an arrangement was discussed with Mr. Mecom at the time of the Milwaukee Lotus Race on August 18. Mr. Mecom seemed amenable to such an arrangement, providing:

1. That we would agree to sell him the first Ford G.T. to be placed in the hands of a private owner.

2. That we would arrange with Colin Chapman to accelerate delivery of a Lotus 19B which had been tentatively ordered by Mecom from Chapman with the vehicle to be powered by a Ford 289 high performance engine.

A firm agreement based on these terms was not reached in Milwaukee since we were in no position to commit Colin Chapman, producer of the Lotus 19B, to a firm delivery date. It subsequently developed that the 19B car requires a "dry sump" engine design which our Engine and Foundry Division is now working on. E & F plans to ship one 19B dry sump engine to Chapman for installation in the first Lotus (destined, we understand, for Dan Gurney) by October 1. On this basis, it was apparent that a dry sump 19B for Mecom would not be available until quite late this year and then only depending on Chapman's wishes. At the meeting in Milwaukee, Mr. Mecom was requested to develop the complete details of an agreement which he would like to enter into with us.

On August 30, the writer was contacted by Tom Tierney, Southwestern Regional Public Relations Manager, indicating that, if Mecom were to sell us the Lola, he would want the Ford G.T. in April or May of 1964 in order to have the car during a portion of the 1964 racing season (Mr. Tierney having been contacted by John Mecom on this point). Mr. Tierney was advised that we would talk with

This & next two pages: Memo from Ford Motor Co detailing its conversation with John Mecom about his GT Mk 6 coupé.

Ford Motor Company FORD DIVISION

Intra-Company Communication GENERAL OFFICE

- 2 -

Mr. Mecom directly and indicate to him that such a delivery was highly improbable since our own prototype testing would still be underway and that more than likely the Ford G.T. would not be able to reach him until much later in the year. At this point, I talked with Mr. Mecom and indicated this to him at which time he said that he felt Ford had interceded with Lola and was, in fact, delaying shipment of the Lola Car to him. We advised that this was not true and called England to ask them to expedite shipment of the Lola to Mecom since we could not reach agreement satisfactory to Mecom with respect to G.T. delivery date. We also conveyed that, while we had no firm control of a 19B delivery date, it was apparent that the 19B delivery to him with a dry sump Ford Engine would be delayed to possibly December of this year. As a result of our expediting shipment, the Lola was sent to Mecom in time for testing at the Elkhart, Wisconsin Sports Car Race.

On Monday, September 9, immediately following the Elkhart Race, Mr. Mecom contacted this office and asked the status of our negotiations. We stated that we had expedited shipment of the Lola to him in keeping with his desire to have the car on hand for Elkhart; we asked what his reaction would be to our providing a 289 high performance engine for installation in the Lola (it now has a Chevrolet engine) and we re-affirmed that, while we would not be purchasing the Lola from him, he would still be one of the first private sponsors to get a Ford G.T. when it is available. Mr. Mecom indicated that he would welcome a high performance 289 engine for installation in the car and steps have been taken to ship this engine to him with the engine to arrive within the next ten days. Mr. Mecom was also advised that I would discuss with Mr. Chapman expediting delivery of a Ford-powered 19B later this year for Mecom.

Following Mr. Jack Roach's call to Mr. Miller, I talked to Mr. Roach by phone, reviewed the background outlined above and was surprised to learn that Mr. Mecom had given Mr. Roach the impression that General Motors was planning to purchase the Lola. While this obviously cannot be verified, it seems surprising that Mr. Mecom would release the car since throughout our entire discussion the point of critical consideration seemed to be the question of having a car to race with. In fact, the point where our discussion broke down stemmed from our inability to "an acceptable delivery date" of the Ford G.T. to Mecom. If, in fact, General Motors is planning to purchase the Lola, it would be my recommendation to re-open discussions with Mr. Mecom and intercept the transaction by buying the car ourselves. Mr. Roach also indicated that Mr. Mecom expressed the desire to secure an aluminum block 1964 Indianapolis Ford engine. I advised Mr. Roach that these engines were in critical short supply and that we would not be able to make a commitment to Mr. Mecom along these lines.

I have endeavored to talk to Mr. Mecom by phone today, but have been unable to reach him as his wife is in the hospital. As soon as we can, we will talk with Mr. Mecom and restate our reasons for not being able to reach an

177

Ford Motor Company   FORD DIVISION

Intra-Company Communication   GENERAL OFFICE

- 3 -

agreement on the purchase of the Lola Car and offer further assistance in the form of providing Ford power plants for his racing cars with the exception of committing a 1964 Indianapolis Engine.

Frank E. Zimmerman Jr.
Frank E. Zimmerman, Jr.

177

The prototype and LGT/1 Lola Mk VI, being worked on in the new Lola factory in Slough, late 1963. (Courtesy Ford Archives)

The Chevrolet Corvette 327in^3 engine was always seen as *the* engine to insert into the Lola GT, being more powerful than the Ford 289. Indeed, John Mecom's car, which Augie Pabst later drove, had such a unit fitted, tuned by Traco in California. Augie Pabst, in an interview with Dave Friedman, later said:

"The Ford 289 was a good engine, but a well-tuned Chevy was undoubtedly more powerful! I remember that Eric Broadley took me for a ride in the car on the street before I drove it at Brands. He scared the shit out of me because it was raining, and we were racing over some very narrow brick roads. Luckily, we didn't hit anything, but you couldn't have convinced me of that at the time. That was one of the few times in my life when I was really scared ... "

John Mecom later recounted that he was summoned to a top secret meeting with Ford, which had learned that he had bought the third Lola GT Mk 6 coupé built. Ford had, by that time, contracted with Eric Broadley to build the GT40 and did not want any of the Mk 6 coupés to go anywhere else except to Ford ...

Indeed, there is a letter extant (reproduced in this

Augie Pabst on the grid at Oakes Field, Nassau, for the start of the Bahamas Tourist Trophy in December 1963. (Courtesy Lola Archives)

1964 at Sebring. The retired Lola GT Mk 6, LGT/2, being towed away by, of all things, a Fiat 500! (Courtesy Lola Archives)

chapter), in which Fred Zimmerman seems to have been quite alarmed at the prospect of General Motors using Mecom's GT Mk 6 as some sort of a test car, but there is no proof that this ever actually took place. Indeed, Augie Pabst later commented on the lack of testing that the Lola suffered from ...

December saw John Mecom entering the Nassau races in the Bahamas, again with Augie Pabst as his driver. Pabst won two races at Oakes Field, a track of 4.5 miles length: the Nassau GT five-lap race, and the Bahamas Tourist Trophy. The latter was a 25-lap race, reduced to 22 laps as darkness was falling toward the end of the race. Pabst was third for several laps, behind the Chevrolet 'Grand Sport' Corvettes of Jim Hall and Dick Thompson. By the 16th lap, Pabst was in the lead and, apart from having to open his door to see, due to his windshield being covered in oil, he had no other problems.

In 1964, Walt Hansgen and Augie Pabst took part in the Sebring 12 Hours with the little coupé, but the engine blew when the ignition timing slipped. Augie Pabst said:

"We didn't do any testing with that Lola. If only we had ... ! It was a great little car, and it's only real problems were with the engine, so if we'd have had a spare weekend to sort the niggling problems, we'd have won a lot more races ... "

The Mecom-entered Lola went to Mosport (engine blew in practice), Road America (engine overheated), The Guards Trophy at Brands Hatch again (finished 11th), Road America again in September (engine problems due to faulty maintenance), and the car was later very badly damaged (it was repaired in later years) at Riverside in October 1964 when it was crashed, again with Augie Pabst at the wheel. During that year, the Lola's wheels (and wheelarches), had steadily grown in width to take the newer, wider wheels and tyres that became available.

The dashboard of LGT/2. (Author's collection)

John Mecom's Lola GT Mk 6 at Road America, before the race in 1964. (Courtesy Gerard Melton)

Chapter 4

The GT and Sports Car Project

In June 1963, Roy Lunn wrote a booklet entitled *GT AND SPORTS CAR PROJECT*. This was distributed to the top directors of the Ford Motor Car Company and those associated with the racing divisions of Ford in order that they could follow the progression and ideas behind the new High Performance and Special Models operation. It was not until after this booklet was issued that Ford engaged Lola Cars Ltd to assist with the design and building of Ford's new GT40, which would debut in 1964.

We are obliged to the Ford Archives for allowing us to use some of this information shown in the following pages.

The Lola GT Mk 6 at the Nürburgring in May 1963. After Ford's delegation to Europe saw it at Le Mans in June, they decided to base the new Ford GT40 on this design. (Courtesy Lola Archives)

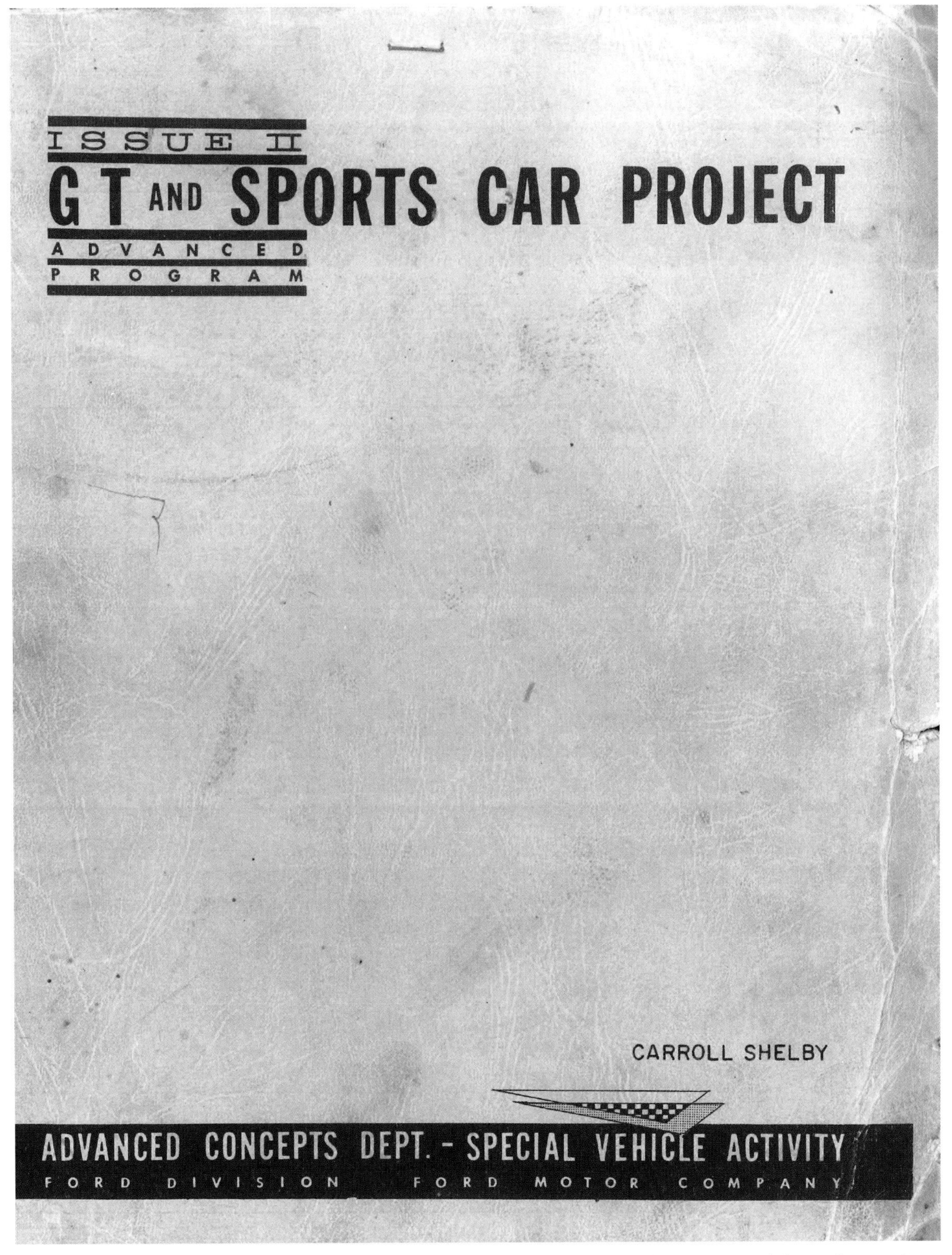
ISSUE II
GT AND SPORTS CAR PROJECT
ADVANCED
PROGRAM
CARROLL SHELBY
ADVANCED CONCEPTS DEPT. - SPECIAL VEHICLE ACTIVITY
FORD DIVISION FORD MOTOR COMPANY

The cover of Roy Lunn's booklet for the Advanced Concepts Department – Special Vehicle Activity of the Ford Motor Company. This one was issued to Carroll Shelby. (Courtesy Ford Archives)

GT AND SPORTS CAR PROJECT

FOREWORD CONFIDENTIAL

In July, 1963, the Special Vehicles Activity was formed within the Ford Division to further the promotion of image building programs. This new activity is to be responsible for promoting exciting and progressive programs which generate in the public's mind a desire to own a Ford product.

One of the new components of this new activity is the Advanced Concepts Department. This group's directive is to develop highly featurized show cars, high performance models, racing cars, and generally keep a watching brief on new developments in the automotive world. The group will take full advantage of advanced company developments, effect liaison with outside specialists where appropriate, and execute vehicle builds, mainly through outside vendor sources.

The first projects selected for establishing the group are:

1. A racing GT car that will have the potential to compete successfully in major road races such as Sebring and Le Mans.

2. A high-performance, two-seater road sports car, in the category of, and superior to, the Corvette. Design will take into account the possibility of its being developed into a low-volume model line.

This package book relates to these two projects and replaces Issue I which was circulated in June, 1963.

The Foreword. It's important to note that, at this stage (July 1963), Ford saw the project as not only being a racing sports car but also as a car to be sold to customers. It was this dual purpose idea that led to Roy Lunn and Eric Broadley falling out, as Lunn wanted a steel chassis for durability of the street version, whereas Eric Broadley could see no use for that in a race car, where a lighter chassis, made of aluminium, would be far better. (Courtesy Ford Archives)

GT AND SPORTS CAR PROJECT

PROJECT HISTORY CONFIDENTIAL

Early in June, 1963, an initial design study was executed by small groups in the Advanced Styling and Research areas: The results were compiled in a package book which served as a basis for discussion with possible vendor sources. It also established the basic objectives of the program. This initial study generated a low, sleek, hard-top package concept, 40 inches high, using a mid-ship engine configuration. The design took advantage of many of the features and ideas developed in the Mustang program, and made provision for installation of the "Indianapolis" push rod and TOHC engines.

It was decided to use a European source for the execution of the project and a survey of possible vendors was made in the latter part of June, 1963. The program was approved on July 12, 1963, and a selection of vendor made by the end of July, 1963. The Lola Car Company in England was selected for two main reasons: Eric Broadley, who runs the company, had already built prototypes of a similar type GT vehicle using a Ford 289 CID engine: Mr. Broadley was also receptive to working on the design with Ford engineers and to having the vehicle promoted as a Ford GT car.

An arrangement was made with the Lola Car Company to procure two of their existing prototypes for use as component development vehicles. These two cars have been fitted with new suspensions, engines, and cooling systems of Ford origin in order to determine the ingredients for the final Ford GT design. Development testing at the Brand's Hatch, Goodwood, Monza, and Snetterton circuits started late in August and extended through the end of November.

Three design engineers from Engineering Research were loaned to Ford Division to participate in the engineering, design, and development of the car. They were transferred to England by mid-October and are presently working on the final design of the racing car.

Engineering Research has provided services for package studies, suspension analysis, and aerodynamic development.

The new suspension designs, now on the existing Lola vehicles, were established in conjunction with the computer programs developed by Research. Extensive aerodynamic testing was carried out with full-sized and 3/8 scale models at the Dearborn and Maryland Wind Tunnels. The results of these tests evolved a low-drag shape and stabilizing lift factors, aimed at controlling the car at speeds in excess of 200 miles per hour.

Advanced Styling areas have provided full-sized and 3/8 models developed in conjunction with package studies, wind tunnel testing, and track evaluation. The initial models were completed early in July and the final shape was shipped to England early in November. This final full-sized scale model is being used for the manufacture of molds and forms to be used in the fabrication of the race car bodies.

Advanced Engineering areas of Engine and Foundry Division have provided two "Indianapolis" push rod engines for initial development testing in the existing Lola vehicles. The results

The project history: Note that in this paper, the use of the Ford 'Indianapolis' twin overhead camshaft engine was envisaged, and two were supplied to Lola. Also, the purchase by Ford of two of the Lola GT Mk 6s is mentioned, as are the tests of the Lola that took place from August to November 1963. Note that mention is made of the Ford-designed suspension, which apparently Eric Broadley did not like, and also the newly designed body shape. When Eric Broadley saw that, he told the Ford engineers that the car would fly when approaching its maximum speed. He was proved right at the tests at Le Mans in 1964. (Courtesy Ford Archives)

GT AND SPORTS CAR PROJECT

PROJECT HISTORY CONFIDENTIAL

of this testing has established the requirement for adapting the "Indy" engine for GT racing. A total of seven push rod engines and two TOHC engines will be required for the projected 1964 race program. Engine and Foundry, in conjunction with General Products Division, also provided the services of an engineer to make the initial electrical installation in the Lola test vehicle.

The Shelby American organization sent a senior engineer to England to install the "Indy" engine. This engineer also acted as chief mechanic while the new chassis components were being installed and while the vehicles were on test.

Transmission and Chassis Advanced Engineering has made a study of a transaxle for application to the GT vehicle. The unit under consideration is a six-speed, semi-automatic transmission which would replace the four-speed Colloti box now being used. The successful development of a unit of this type would increase the Ford content in the car and diminish driver error.

The original Lola factory was located in Bromley, Kent, on the southeast side of London. This facility was not large enough to meet the program requirements and additional labor was difficult to find in the Bromley area. A new factory has been rented at Slough, Buckinghamshire, on the west side of London, which also is the labor center for the type of personnel required. The new factory was started at the beginning of November and the facility made operable by the end of November. Mr. Wyer, former Aston Martin General Manager, was engaged at the beginning of September. He is handling the administrative and commercial aspects of the GT program, together with liaisoning other projects being enacted in Europe. He will also be responsible for organizing the racing program and managing its associated aspects. When the Lola facility moved to Slough there were only five of its original staff remaining. Three additional mechanics have been engaged, together with an accountant and odd-job man. A further eight personnel are now being sought.

Major vendors are being selected for the supply of major chassis and body items. These include a sheet metal company for making the unitized underbody; a plastics company for the manufacture of molds and outer-skin panels; also companies for providing wheels, tires, brakes, clutches, transmissions, lamps, etc. A design company is also being used for temporary design help.

Work thus far on the project has mainly been confined to the GT race car. The road sports car will be derived from this model and work will commence on it as soon as the basic design of the race car is crystallized. This issue of the package book relates mainly to the GT racing car.

The project history (continued). This paper gives the information that Shelby American sent one of its engineers (Phil Remington), to England to oversee the installation of at least one Ford 'Indianapolis' V8 engine. He also assisted with the testing. It was also envisaged that a six-speed semi-automatic gearbox would be used at this time. Mention is also made of the new Ford-supplied Lola factory in Slough, which started work in November 1963. New engineers were urgently needed. September saw John Wyer being taken on to oversee the racing and commercial departments. At this time, the race cars were top priority. (Courtesy Ford Archives)

GT AND SPORTS CAR PROJECT

PACKAGE DESCRIPTIONS CONFIDENTIAL

CONFIGURATION

The two vehicles under consideration are the GT road racing car and the road sports version.

The GT vehicle is a high-performance, two-passenger sports car, designed for competition in the Grand Touring Prototype Class in accordance with the Federation Internationale de la Automobile (F.I.A.), Appendix J, regulations. It is a mid-ship engined, rear-drive vehicle, utilizing the "Indianapolis" Fairlane V-8 engine with a 256 cu. in. displacement.

The aerodynamic, fast-back coupe body presents a silhouette of 40.5 inches in height to the top of the roof.

Its capabilities range from straightaway speeds in excess of 200 mph to tight-corner maneuvering. With these objectives in mind, a 95-inch wheelbase was chosen to present a minimum package for optimum power-to-weight ratio and handling response.

The road sports car is a vehicle which is equally "at home" on the road or the race course; however, since its basic environment is that of a road vehicle, certain compromises are made. These are found in areas such as body construction, suspension and engine noise isolation, and overall height of 43-46 inches. This vehicle uses the cast iron, high-performance, 289 cu. in. Fairlane engine.

Consideration has been given in both packages for the accommodation of the O.H.C. version of the Fairlane engine.

Both vehicles use the mid-ship engine and basically the same chassis equipment. The cars also have a family relationship in their styling, although the road car is higher and will be fitted with normal highway equipment such as bumpers.

BODY AND FRAME

The main structure on the GT car is semi-monocoque steel construction, utilizing the roof skin section as a stressed member. Pontoon sections form the sills, and serve as the basic structural members through the mid-section. Sheet steel members extend fore and aft, providing attachment points for engine and suspension.

Entry to the cockpit is provided by means of a front-hinged door opening into the roof.

The seats are fixed and form an integral part of the chassis, as does the roll bar.

Functional aerodynamic form is the basis of the body shape design. The body has been styled to low air drag requirements with the combination of small frontal area and low drag coefficient.

Package descriptions (P1): A technical description of the forthcoming Ford GT40 race car, using the Ford 'Indianapolis' V8 engine. (Courtesy Ford Archives)

GT AND SPORTS CAR PROJECT

PACKAGE DESCRIPTIONS CONFIDENTIAL

The road sports car structure and type of door opening has yet to be determined. Feasibility of design with regard to production facilities must be analyzed.

Many features and components are common to both vehicles. These include "through-flow" ventilation, retractable headlights, fixed seats, movable controls, ventilated seats, lumbar support seats, retractable seat belts, and aerodynamic cooling system.

ENGINE

The engine for the GT car is an aluminum "Indianapolis" 256 cu. in. Fairlane V-8, mounted in the mid-ship position forward of the rear axle.

It develops well over 350 horsepower in the 6000 to 8000 rpm range on gasoline fed through four dual-throat, 48 mm, Weber downdraft carburetors. Provisions in the chassis design have been made for the O.H.C., V-8 engine which is presently in the development stage.

The basic engine for the road sports car is a 289 cu. in., high-performance Fairlane unit in cast iron. Other higher performance engine options will be available, up to and including the "Indianapolis" engine. Air scoops above rear wheel housing provide clean, cool air to carburetors.

TRANSAXLE AND CLUTCH

The unit being used for the 1964 racing season is of Colloti manufacture. It incorporates a manual, four-speed, non-synchromesh transmission and limited slip differential unit. The clutch is an 8-1/2 inch diameter, twin-plate Borg and Beck unit which is hydraulically operated.

A study of a six-speed automatic unit is being made by Transmission and Chassis Division. This unit comprises of an XP3 automatic box and a "splitter" arrangement. Selection will be manual with a clutch device for starting from rest.

A five-speed synchronized manual transmission is being designed by ZF in Germany for possible application on the 1965 cars and the production models.

FRONT SUSPENSION

The front suspension units are common to both the GT and road version. They are independent, double-wishbone type. Both upper and lower arms are of welded tubular construction. Their attachment points are well spread to distribute the loads, and are adjustable. Anti-dive is a feature of the design. Adjustable coil-spring shock units are employed.

Package descriptions (P2): Further technical details. The 'Indianapolis' V8 engine was expected to develop 350 horsepower between 6000-8000rpm. The road version of the race car would use the 289in³ pushrod engine. The gearbox for the race version in 1964 would still be the old Colotti unit, but ZF was designing a five-speed and reverse gearbox for the car. The suspension is detailed. (Courtesy Ford Archives)

GT AND SPORTS CAR PROJECT

PACKAGE DESCRIPTIONS CONFIDENTIAL

REAR SUSPENSION

The rear suspension, just as the front, is common to both vehicles. A full-independent system is employed with an objective to fulfill the ideal tire-to-road relationship under all conditions for maximum cornering ability.

The arrangement comprises of two trailing links in combination with an inverted lower A-arm and a single-strut upper arm. Coil-spring shock units are used.

Anti-lift and anti-squat have been designed into the geometry of the system to control the attitude of the vehicle.

Being adjustable, the suspension may be tailored for the particular handling requirements of any road course.

Attachment points of both the front and rear suspensions on the GT car make use of low friction metal joints, while the road sports car will combine these with rubber isolators.

BRAKES

The front and rear wheels are equipped with 11-1/2 inch Girling disc brakes. A dual master-cylinder arrangement provides for separate front and rear systems. This dual arrangement, in addition to providing a safety feature, permits tailoring of the braking distribution.

STEERING SYSTEM

The steering gear is a forward-mounted, rack-and-pinion unit offering direct, responsive control. The rack is fitted with an external hydraulic damper.

CONTROLS

The seats are an integral part of the structure and the clutch, brake, and accelerator pedals are mounted as an adjustable unit to accommodate various driver sizes.

WHEELS

Wheels are 15-inch wire type with aluminum rims and Rudge hubs.

TIRES

Dunlop has designed special tires for the vehicle which are 5.50 x 15 on the front, and 7.25 x 15 on the rear.

Package descriptions (P3): This details the rear suspension, the Girling disc brakes, steering type, the fixed seats (the pedals were adjustable for the length of the driver's legs), wheels and tyres. (Courtesy Ford Archives)

GT AND SPORTS CAR PROJECT

PACKAGE DESCRIPTIONS CONFIDENTIAL

FUEL SYSTEM

On the GT car, two fuel tanks of aircraft fuel-cell design are used and are fitted in the sills on both sides of the vehicle. They provide for a total capacity of 34 imperial gallons. Large-diameter filler necks to both tanks permit a quick-fill operation.

COOLING SYSTEM

A light, high-efficiency, cross-flow radiator is mounted in the front end of the car. The ducting is arranged to form an aerodynamic system where air is taken in at a high-pressure area and extracted in a low-pressure area.

EXHAUST SYSTEM

The exhaust system is of the tuned type for maximum scavenge effect. Each port exhausts to its own pipe. These pipes then join and form two short exhaust extensions to which silencers are fitted.

INSTRUMENTATION

The instrumentation includes a mechanically driven tachometer, oil temperature and pressure gauges, as well as ammeter, fuel pressure, and water temperature gauges.

HEAD LAMPS

The initial prototypes will use CIBIE rectangular lamps, rigidly fixed and fully shrouded for aerodynamic efficiency. Subsequent installations will feature retractable elements as an improvement in overall efficiency and to meet regulations.

Package descriptions (P4): This covered the fuel system used, the cooling and exhaust systems, instrumentation and headlamps. (Courtesy Ford Archives)

GT AND SPORTS CAR PROJECT

VEHICLE SPECIFICATIONS - GT PROTOTYPE CONFIDENTIAL

		Original Lola	Reworked Lola	Ford GT Prototype
GENERAL				
Wheelbase	(in.)	92	92	95
Tread				
Front	(in.)	51.5	52	54
Rear	(in.)	51.5	52	54
Overall				
Length	(in.)	153.8	153.8	158.6
Width	(in.)	61.5	68.2(over scoops)	70 (over scoops)
Height	(in.)	40	40	40.5
Overhang				
Front	(in.)	36.4	36.4	35.5
Rear	(in.)	25.4	25.4	28.7
Approach Angle	(deg.)	15	15	15
Departure Angle	(deg.)	22.5	22.5	32
Miscellaneous Height				
Cowl	(in.)	27.4	27.4	28.25
Windshield	(in.)	38.1	38.1	39.2
Steering Wheel Top	(in.)	30.8	30.8	31.35
Minimum Ground Clearance	(in.)	4.8	4.8	4.8
Weights				
Basic (No Fuel)	(lbs.)	1864	1784	1779*
Front Distribution	(lbs.)	784	804	745
Front Distribution	(%)	42	45	42

*As of December 1, 1963, regulation 4.2 liter weight is 1914 lbs.
This minimum weight is to be achieved through ballast.

BODY

The GT car employs monocoque construction of .024 steel with sub-frame extensions and the technique of welding structure to outside surfaces. Hinged front and rear panel sections and doors are of reinforced fiberglass.

The structure for the road car has yet to be determined.

Details the vehicle specifications, showing the differences between the Lola GT Mk 6 and the projected Ford GT40. (Courtesy Ford Archives)

GT AND SPORTS CAR PROJECT

VEHICLE SPECIFICATIONS – GT PROTOTYPE CONFIDENTIAL

		Original Lola	Reworked Lola	Ford GT Prototype
GT ENGINE				
Cylinders				
Number		8	8	8
Bore	(in.)	4.00	3.76	3.76
Stroke	(in.)	2.87	2.87	2.87
Displacement	(cu.in.)	289	255	255
Compression Ratio		11.6	12.5	12.5
Power Output (Dynamometer)				
Maximum BHP	(@ rpm)	312 at 6000	376.4 at 7200	376.4 at 7200
Maximum Torque	(@ rpm)	271 at 3400	294 at 5600	294 at 5600
Carburetors		4 Dual Barrel 48 mm Webers	4 Dual Barrel 48 mm Webers	4 Dual Barrel 48 mm Webers
Lubrication System				
Capacity	(qts.)	5	15	15
TRANSAXLE				
Ratios				
Axle		3.09 - 3.55	3.09 - 3.55	3.09 - 3.55

TRANSMISSION

Colotti		Optional Fourth Gears	Optional Third Gear
First Gear	2.50	1.07	1.35
Second Gear	1.70	1.10	
Third Gear	1.29	1.12	
Fourth Gear	1.00	1.17	
Reverse	2.50	1.20	
		1.23	

NOTE: Same transmission used on all three cars.

CLUTCH				
Two Discs				
Diameter	(in.)	8.5	8.5	8.5

Further vehicle specifications. (Courtesy Ford Archives)

GT AND SPORTS CAR PROJECT

VEHICLE SPECIFICATIONS – GT PROTOTYPE CONFIDENTIAL

		Original Lola	Reworked Lola	Ford GT Prototype
FRONT SUSPENSION				
Wheel Travel				
Jounce	(in.)	3.0	3.0	3.0
Rebound	(in.)	3.0	3.0	3.0
Camber				
Design	(deg. negative)	1.0	0	1.0
Rate of Change	(deg./in. wheel travel)	1.0	.7	.7
Caster	(deg. positive)	6.00	6.00	6.00
Toe In	(design in.)	.06	.06	.06
King Pin Inclination	(deg.)	8.0	8.0	8.5
Scrub Radius	(in.)	1.75	1.75	2.00
Roll Center Height	(in.)	5.00	4.10	4.10
Anti-Dive	(%)	25	75	54
REAR SUSPENSION				
Wheel Travel				
Jounce	(in.)	3.0	3.0	3.5
Rebound	(in.)	2.0	3.0	3.0
Camber				
Design	(deg. negative)	1.50	2.00	2.00
Rate of Change	(deg./in. wheel travel)	.84	1.20	1.16
Toe				
Jounce	(deg. in)		.03	.01
Design	(deg. in)	N.A.	0	0
Rebound	(deg. out)		.12	.08
Roll Center Height	(in.)	3.0	4.60	4.64
Anti-Lift	(%)	0	68	54
Anti-Squat	(%)	0	60	48

Further vehicle specifications. (Courtesy Ford Archives)

GT AND SPORTS CAR PROJECT

VEHICLE SPECIFICATIONS – GT PROTOTYPE CONFIDENTIAL

		Orginal Lola	Reworked Lola	Ford GT Prototype
BRAKES				
Front Brake				
Disc Diameter	(in.)	11.0	11.5	11.5
Caliper		Girling "BR"	Girling "CR"	Girling "CR"
Rear Brake				
Disc Diameter	(in.)	11. 0	11.5	11.5
Caliper		Girling "BR"	Girling "BR"	Girling "BR"
STEERING SYSTEM				
Steering Ratio	(overall)	14.0	14.0	14.0 (Optional 20.0)
Turns (lock-to-lock)		2.8	2.8	2.8
Inner Wheel Turn	(deg.)	25	25	27
Outer Wheel Turn	(deg.)	25	25	27
Turning Circle Diameter	(ft.)	42	42	40
Steering Wheel Diameter	(in.)	15	15	15
Steering Wheel Adjustment	(in.)	2.0	2.0	2.0
GT WHEELS AND TIRES				
Wheels - Size				
Front		6.50 x 15	6.50 x 15	6.50 RIM x 15
Rear		8.00 x 15	8.00 x 15	8.00 RIM x 15
Tires - Size				
Front		5.50 x 15	5.50 x 15	5.50 x 15
Rear		7.00 x 15	7.00 x 15	7.25 x 15
Tires - Loaded Radius				
Front	(35 psi)	13.00	13.00	13.00
Rear	(35 psi)	14.18	14.18	14.25
FUEL SYSTEM				
Tank Capacity	(imperial gallons)	32	32	34

Further vehicle specifications. (Courtesy Ford Archives)

GT AND SPORTS CAR PROJECT

VEHICLE SPECIFICATIONS - GT PROTOTYPE CONFIDENTIAL

		Original Lola	Reworked Lola	Frod GT Prototype
COOLING SYSTEM				
Front-Mounted Radiator - Cross Flow				
Total Area	(sq. in.)	294	294 (Top Exhaust)	294 (Top Exhaust)
Depth	(in.)	5.0	5.0	5.0
Fins per inch		12	12	12
EXHAUST SYSTEM				
Type		Two Separate Banks	Tuned	Tuned
Tube Diameter	(in.)	1.5	1.5	1.5
ELECTRICAL SYSTEM				
System		12	12	12

Further vehicle specifications. (Courtesy Ford Archives)

The Ford 289 engine with Weber carburettors. (Courtesy Ford Archives)

The Ford four-camshaft 'Indianapolis' engine. It's probable that, after testing, this engine was not used, as it produced its power at the top of the rev range. Fine for oval track racing but not for a road course, where torque and horsepower were needed to accelerate away from corners. (Courtesy Ford Archives)

The Lola GT Mk 6 being tested at Goodwood. (Courtesy Ford Archives)

The Lola GT Mk 6 as it was tested at Monza. Note the additional air intakes at the rear, and the cut-aways in the nose, presumably to let hot air escape from the radiator. (Courtesy Ford Archives)

The Ford 289 engine, as installed in the Lola GT Mk 6 in testing. (Courtesy Ford Archives)

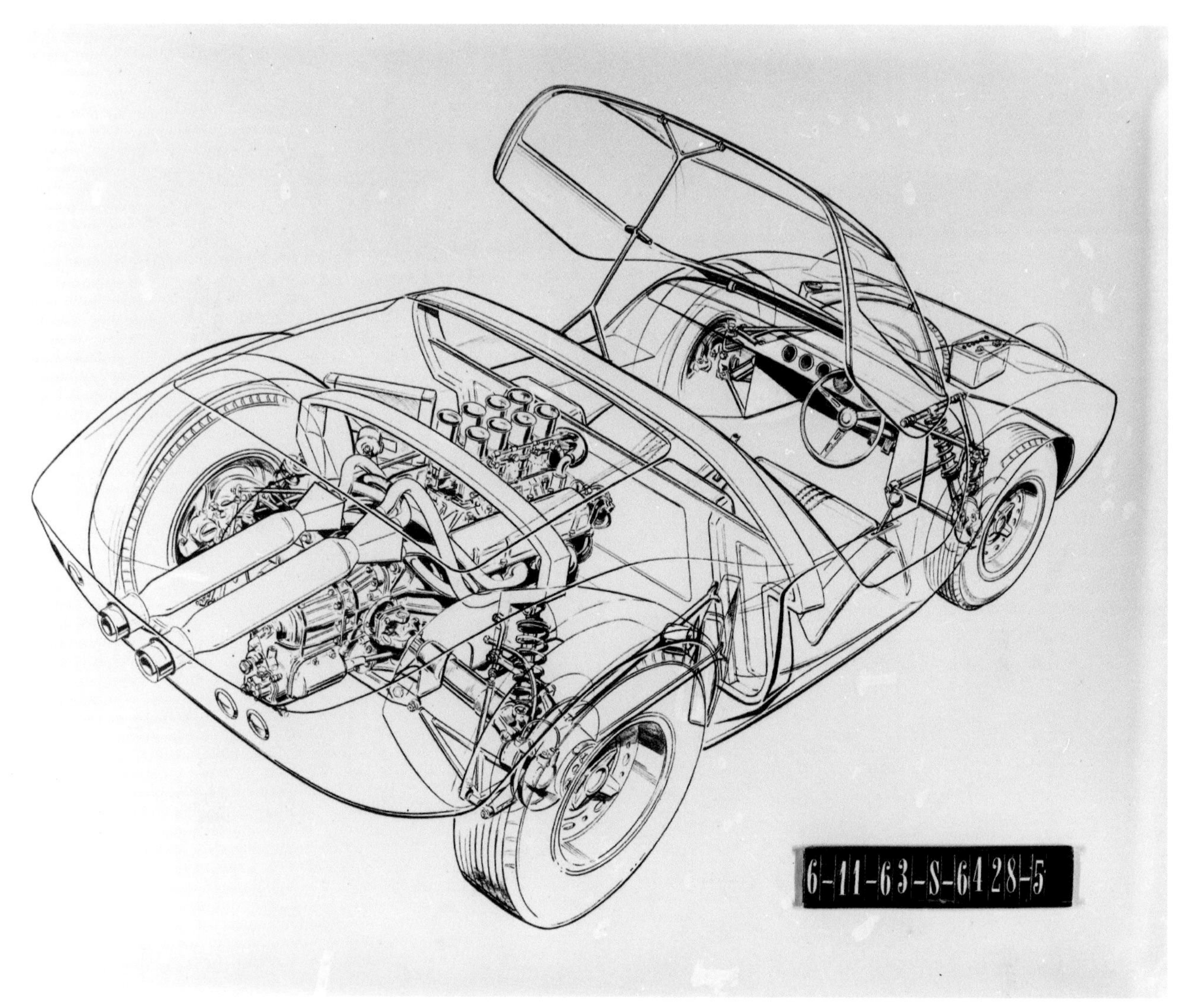

One of the initial Ford drawings for the projected Ford GT car. This one had the lift-up front, which, as Eric Broadley pointed out to Ford, "Wouldn't work. If the car turned over in an accident, the driver wouldn't be able to get out ..." (Courtesy Ford Archives)

An artist's impression of what the new Ford GT would look like. (Courtesy Ford Archives)

Drawings of a projected Ford GT from the first edition of GT AND SPORTS CAR PROJECT dated 11th June 1963.

FORD

This page & following pages: Various views, dated 23rd October 1963, of the new bodywork for the forthcoming GT40. (Courtesy Ford Archives)

FORD
10-23-63
S-6775-8

7
FORD
10-23-63
S-6775-3

FORD
10-23-63
S-6775-4

10-23-63
S-6775-5

Chapter 5

Ford beckons

On the other side of the Atlantic, machinations were occurring that would impact Lola Cars Ltd greatly in the coming years.

In 1960, John Kennedy became the President of the United States. For his secretary of defence, he selected Robert McNamara, at that time the General Manager of the Ford Motor Company. In his place, Henry Ford II appointed Lee Iacocca. Iacocca wasted no time in introducing a new programme, aimed at the younger people of America, in an attempt to rid Ford of its 'Fuddy Duddy,' old fashioned image. At that time, Fords were what most school and college kids' parents drove. The image, in other words, was staid and lacklustre. Iacocca set out to change that.

He instituted marketing meetings and they came up with the phrase 'Total Performance.' It was to prove a strong message in the next few years, and one of its results was the introduction of the Ford Mustang in 1964, which gave Ford a huge sales success. Realizing that the younger customers wanted the excitement that racing could provide, and wanting the glamour that was attached to auto racing in Europe, Ford looked around to find, and to win, the most prestigious race there ...

Iacocca and his marketing men decided to expand Ford's GT racing operations from its USA homeland in NASCAR to Europe. They reasoned that the publicity so gained would help appeal to younger buyers of their 'bread and butter' cars.

Because the Le Mans 24 Hours race was *the* race to win in order to establish its new performance-oriented image with the car-buying public in Europe and America, Ford also realized that it lacked the racing knowledge that its European counterparts

Lee Iacocca. (Courtesy Historic Images Outlet)

Enzo Ferrari in 1984. (Author's collection)

Roy Lunn. (Courtesy the late Graham Robson)

possessed of long distance racing sports cars, so Ford set out to buy that knowledge.

Specifically, Ford, through his representative Don Frey, approached Enzo Ferrari in 1963 with a view to purchasing his company. Ferrari was not averse to this offer: he was, after all, no longer a young man, and had been financing his company's racing for several years by building expensive road cars to sell to a small, well-heeled sector of the public. Naturally, this involved extra effort, an effort he begrudged, having a passion only for true racing cars; that is, those cars that would bring him and his factory glory on race tracks around the world. Making road-going cars, no matter how glamorous they were, was simply a means of financing his racing passion.

Initially, the asking price for his company to sell to Ford was $18 million, but, as Ford accountants scoured the details of his books, the amount offered became lower and lower until, realizing just how little autonomy he would have left after the deal, specifically with regard to his beloved race cars, Ferrari broke off the negotiations on 21st May 1963.

Lee Iacocca was not disillusioned. He simply decided to buy other European expertise and create a totally Ford-controlled project. To do this, he knew that he would have to hire one of the pre-eminent English racing chassis builders. He sent Roy Lunn, an ex-English engineer, who had worked in the car industry in Britain, first of all as a designer for AC Cars, then Jowett, and later Aston Martin. Ray Geddes, Hal Sperlich and Carroll Shelby accompanied Lunn to Le Mans in 1963, where they undoubtedly saw and were interested in the Lola GT Mk 6.

However, it has now become apparent that Ford was interested in the Lola GT Mk 6 before Le Mans, which usually takes place around the third weekend in June. Roy Lunn brought out the first edition of his booklet *GT and Sports Car Project* for distribution internally to top managers at Ford on 12th June 1963. It is obvious in reading this book that, although not actually mentioned, Ford management must have been talking already to Lola about the Lola GT

Just arrived; The first production Lola GT Mk 6 at Le Mans in June 1963. Hours of work awaited the crew to get it ready for the start. (Courtesy Lola Archives)

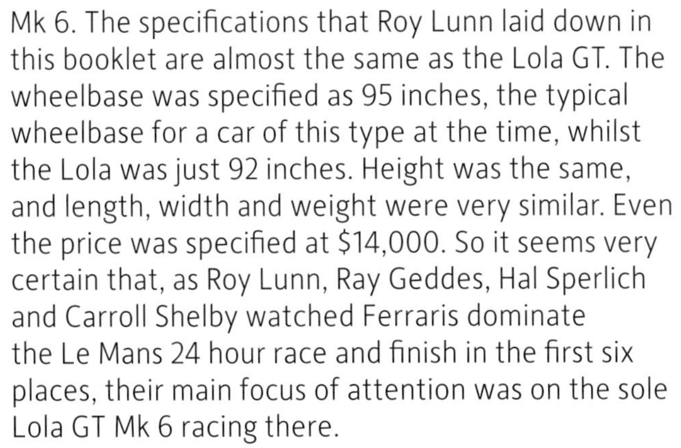

Mk 6. The specifications that Roy Lunn laid down in this booklet are almost the same as the Lola GT. The wheelbase was specified as 95 inches, the typical wheelbase for a car of this type at the time, whilst the Lola was just 92 inches. Height was the same, and length, width and weight were very similar. Even the price was specified at $14,000. So it seems very certain that, as Roy Lunn, Ray Geddes, Hal Sperlich and Carroll Shelby watched Ferraris dominate the Le Mans 24 hour race and finish in the first six places, their main focus of attention was on the sole Lola GT Mk 6 racing there.

The quartet had flown to England before Le Mans, and had visited the Cooper, Lotus and Lola companies. They decided that Cooper, whilst having been successful in F1 up to 1960, had fallen back in the technology stakes. They were wary of Colin Chapman, the boss of Lotus, with whom they had already worked on the project to win the Indianapolis 500, and considered that Chapman would want total control of the whole project. In fact, at their meeting with Colin Chapman, he offered to sell them his entire Lotus company, less the racing side. Lunn, Geddes and Shelby were not interested in buying a factory producing sports cars for the street, however, and they turned him down.

The new Ford-designed bodywork, here in mock-up form, and probably in fibreglass, mounted on a Lola GT Mk 6's chassis. The photos were taken on 4th October 1963. (Courtesy Ford Archives)

Their attention then turned to Eric Broadley and Lola Cars Limited. When the Ford delegation visited

Ford promotion photos taken on 12th July 1963, show three views of the mocked-up new Ford GT car compared to its two main rivals, the Jaguar E Type/XKE and the newly introduced Chevrolet Corvette. When compared to the E Type/XKE and the Corvette, the Ford appeared very small. (Courtesy Ford Archives)

Front and rear 3/4 views of the proposed new Ford GT car: photos also taken on 12th July 1963. (Courtesy Ford Archives)

the factory in Bromley, they saw the operation and the GT Mk 6 in detail. At first, they were shocked that Lola's employees worked in such cramped quarters, but were very impressed with the GT Mk 6, realizing that they could use its design to give Ford the weapon that it needed to win at Le Mans. Lunn in particular saw that the Lola GT Mk 6 car was the perfect car on which to base the new Ford GT race car.

On the Ford quartet's return to Dearborn, Lunn recommended that Ford should do a deal with Lola Cars Ltd, and Ford started negotiating a 12-month contract, later extended, with Eric Broadley, to produce a new car based upon the Lola GT Mk 6, to be called the 'Ford GT.' It was to be managed from America by what was called 'Advanced Concepts Department – Special Vehicle Activity,' of which Lunn was the manager. This was formed in July 1963 and Lunn was the Project Manager, soon to be based back in England, but reporting to Frank Zimmerman in Dearborn.

All this led directly to Ford entering into an agreement with Eric Broadley as a 'Design Consultant,' and using Lola Cars Ltd to design and build the forthcoming Ford GT40, in conjunction with Roy Lunn, an arrangement which, as the project progressed, Broadley had problems with, as well as with the Ford GT itself.

Roy Lunn was seen by Ford to be the right man to head up the new deal that had been agreed with Eric Broadley, particularly as he had previously lived in England and was an engineer himself. He had played a part in the development of Aston Martin's DB2 entry at Le Mans in 1950. After leaving the RAF, where he had been a pilot, Lunn went to work at AC Cars of Thames Ditton in 1946. In 1949, he worked for Jowett, becoming the chief designer there. Then, after a spell at Aston Martin, he went to the Ford Car Company, also in the UK, starting a research centre in Birmingham, England. He emigrated to America in 1962 and went to work for Ford. There, he helped to develop the Mustang I, a mid-engined prototype car that bore few similarities to the Mustang road car that was eventually introduced to the public in 1964.

According to John Wyer, in his book *The Certain Sound*, in July 1963 Eric Broadley signalled an

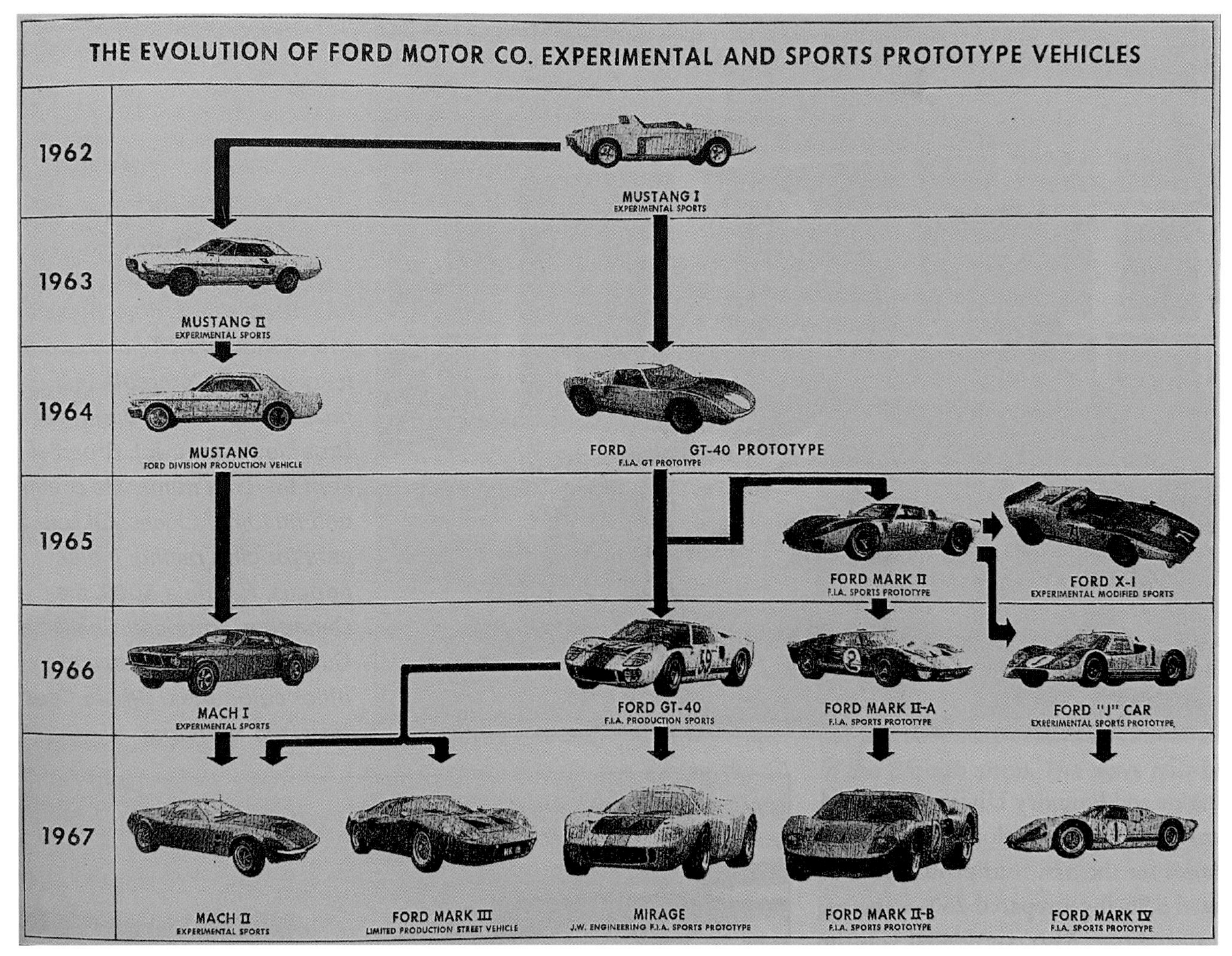

A chart produced by Ford, showing the lineage of the GT40. Somehow, the Mach 1 Mustang appears to be its forebear. There is no mention of the Lola GT Mk 6! (Courtesy Ford Archives)

John Wyer: the man behind Ford's operations in England, circa 1963-66. (Courtesy the late Graham Robson)

intention to sign a one-year contract with Ford. The contract was drawn up in draft form, but Eric Broadley had some reservations about it and, in the hurly-burly of trying to get the Ford GT40 designed and built, this was not seen as being the most important matter, and was not signed by him until later. On 1st August, Lola Cars Ltd moved from Bromley to its new factory in Slough, next to Ford Advanced Vehicles, where John Wyer was the managing director.

Laurie Bray:

"When the Americans from Ford came over, they were, to put it mildly, astonished to find Eric operating from such a small base, and Roy Lunn invited me and Andy and Terry over to the Palace Tavern across the road from the shop and, over a drink, apologized to us for having to work from such a small place when working for Ford. Soon after that, in November, we moved to the new factory in Slough."

Ford purchased two of the cars (chassis numbers LGT/P and LGT/1), and both underwent several tests at Brands Hatch, Goodwood and Snetterton in the hands of Eric Broadley, Roy Salvadori and Bruce McLaren, who was most enthusiastic about the little coupé. The car was also tested at Monza in October of 1963.

The first test was at Brands Hatch on 31st August. It was a standard test session, with other cars also using the track. According to Roy Lunn's notes, two GT Mk 6s were taken. Lunn's notes say that No 1 was the Chevrolet-engined car of John Mecom (LGT/2), and the No 2 was the Cobra Fairlane-powered version the prototype LGT/P. Eric Broadley drove the car(s) and the No 2 car was used the most. This posted a best lap time of 1 minute 4 seconds in the first test session of four laps. The No 2 Lola was then used again, and posted a best lap time of 1 minute 0.5 seconds over seven laps. It was then the turn of the No 1 Lola, but that only did three laps before the team swapped back to using the No 2 car, and with this Eric got down to a best of 59.5 seconds before a last session with the No 1 car, which posted a best lap time of 1 minute 1 second.

Before the test, the front springs of the No 2 Lola were uprated and, during the test, the damper settings were increased. In conclusion, Eric Broadley's comments were that:

"The car ... was felt to have more of a pitching motion, with its centre of oscillation being about three feet in front of the vehicle. This could be due to the increased spring rate, but no real conclusion can be drawn here due to the unknown factor of the rubber bushing rates in the suspension attachments."

For the second test, the crew moved to Goodwood on 9th October 1963. This was seen as a faster circuit than Brands Hatch. Only one car was taken this time, noted as the "modified" Lola GT (LGT/2). It had the Indianapolis Fairlane engine installed and the suspension front and rear was noted as being a "Ford modification." It was noted in the preamble to the test report that the GT lap record was then 1 minute 28.4 seconds, accomplished by Graham Hill, driving a lightweight E-Type Jaguar.

1964, and Lola Cars Ltd and Ford Advanced Vehicles at last have their own separate factories. (Courtesy Allen Grant)

The prototype GT Mk 6 being tested in late 1963 at the Goodwood circuit. (Courtesy Allen Grant)

Roy Lunn, Bruce McLaren and John Wyer look pensive as they stand behind the Lola GT Mk 6 that they are testing at Monza in 1963. (Courtesy Ford Archives)

Once again, Eric Broadley did the driving, on a slightly damp track. All the suspension settings were noted before the start of the test. The report went on to say:

"The car was felt to be improperly geared for this course. Second gear was used almost entirely(!), with only one opportunity to use third gear. A rev limit of 5000rpm was used. Eric came in after four laps to point out that the gearbox was playing up, the car sticking in second gear, the front wheels were out of balance and the 58mm Weber carburettors were not running cleanly in the next session, after another four laps, Eric called into the pits to report that the rev counter (tachometer), was in error: despite this, he reported that the car: "Felt quite good."

Numerous small problems were noted (gear linkage, throttle linkage, engine moving around excessively), but in the afternoon, Eric got down to a 1 minute 36.2 seconds best time. At 3:28pm Bruce McLaren replaced Eric in the driving seat but only managed just under two laps before pulling off 100 feet before the pits, because of a heavy vibration, due to a halfshaft beginning to fail. What with that and a constant misfire, it was obvious that more testing was necessary.

Four days later, the third test session was again at Goodwood, and once again Eric Broadley and Bruce McLaren drove the same car, LGT/2, as in the second test. No fewer than 14 alterations had taken place on the Lola, including:

New rear engine mounts installed
Electrical and mechanical Tachos installed
New halfshaft inner flanges and reinforcing plates installed and the gearshift mechanism reworked

Eric drove the first five laps to warm up the car and came in to report that neither the electrical nor the mechanical tachometers were working properly. After another nine laps, during which Eric got down to 1 minute 35.8, he came back in, and then, with Peter Jackson of Specialized Moldings in the passenger seat, he attempted to find a fuel leak (which he succeeded in doing – it was in the balance tube). Bruce McLaren then replaced him in the driver's seat, and in five laps got down to a 1 minute 29 seconds time. He came into the pits to say that the throttle linkage was "Too touchy" and that the car was oversteering slightly. A front camber change to 0 degrees was tried, and Bruce went out again and got down to 1 minute 27.3. There went Graham Hill's lap record ...

After lunch, Bruce carried on and finally posted several laps at 1 minute 26.8. More modifications to the suspension settings were carried out and, after a total of some 56 laps, Bruce McLaren left the lap record at 1 minute 25.8, almost three seconds below the official lap record. Eric then took over but came in on lap 60 to say that the seat adjustment bar had dropped out of place. The test came to an end with a note to say that both the front and rear spring rates should be increased from 257psi at the front, to 325psi, and at the rear from 188psi to 205psi.

Some three days later, on 16th October, the team was back at Goodwood for test number four, and this time, the No 2 car (LGT/1), was taken, as was the 'prototype' car (LGT.01) fitted with a Cobra Fairlane engine. LGT/1 was fitted with Hillborn fuel injection. Eric Broadley and Bruce McLaren again did the driving. Conditions were bad, very wet and raining hard. Eric, as usual, drove for the first 12 laps but came in three times to say that the engine wasn't running properly. Various adjustments to the fuel mixture were tried, and then, shortly after midday, Bruce McLaren took over the driving. He also complained about the fuel mixture, and this improved when the injection system's bypass valve was drilled out. At around 3.30pm on a cold and wet day, McLaren came to say that the "whole difference between the two cars is that the prototype is understeering and the No 2 car is oversteering." At 4.50pm, on lap 40, McLaren came in with car No 2, and the test of that car finished.

Eric and Bruce both drove the prototype in the afternoon, but conditions were so bad that the test was ended. Both drivers commented that the prototype car felt very stable and comfortable in the rain.

Much work back at base was carried out before the team headed back to Goodwood for its last test there, on 21st October 1963. LGT/1 was found to have its wheelbase longer on one side than the other by over half an inch ... A complete re-adjustment to get the rear suspension symmetrical side-to-side was carried out. With this knowledge, it was considered "that any suspension evaluation up to this point must be considered irrelevant."

It was to Monza in Italy next, for a test (number 6), from 29th October to 2nd November, and there the No 2 car (LGT/1) fitted with the Hillborn injection was used. The engine proved very hard to start and would die almost immediately. Bruce McLaren was the sole driver on this occasion.

Once the engine was running and after warming up the car for four laps, Bruce posted a 1 minute 36.4 seconds, but then came in with the engine overheating. The radiator was found to be only half full. With this remedied, Bruce went out again but kept on coming back in to report the fuel pressure dropping in critical places on the circuit. In the afternoon, Phil Remington bypassed the pump and filter and starting improved. The handling was reported to be "Much better."

Despite the fuel supply problems, Bruce McLaren posted a 1 minute 27.8 second lap at around 3.15pm and then, at 4.30pm, Roy Salvadori took over driving duties. The car proved again to be difficult to start and so, as light rain had started, the expedient was tried of re-routing the fuel lines to a gallon can, held in Phil Remington's lap, whilst Bruce McLaren drove. This appeared to cure the problem, so now the gallon can was taped in place and Bruce McLaren declared that that "Seemed to be solving the problem." The test was concluded at 5.20pm, with Bruce declaring that "the car was much improved from the standpoint of its handling characteristics."

The team was ready to test at Monza again on 31st October, and arrived there at 8.30am. Once again Eric Broadley warmed up the car. After nine laps, Bruce McLaren took over. He did two laps in the 1 minute 45 range and then came in to report, again, about rear end lift, but also that the car was now achieving some 180mph at the end of the pit straight.

1st November found Monza under rain conditions and "very cool." Nothing very much was achieved and the team returned the next day, only to find that the rain had worsened, and with this, the test was called off. At one point, with Eric Broadley driving, the rear bodywork had flown off the car on the pit straight, and so the need for "sound body attachment" was noted ...

The No 2 Lola (GTP/1) had also been taken to Monza, and both Salvadori and McLaren took it in turns to drive it. McLaren finally got its lap time down to a 1 minute 29.4. This car had been fitted with larger brakes, and the emphasis on this test was to find out how much the braking had been improved.

A final test at Snetterton was undertaken on 27th November. No fewer than 21 items were checked

Snetterton for yet another test by Ford in 1963. (Courtesy Lola Archives)

and/or changed on the No 2 car (the only Lola used for this final test). Eric Broadley, Bruce McLaren and Roy Salvadori were the drivers. It was noted in the report that the then current lap record for GT cars, held by Roy Salvadori in an Aston Martin DB-4 GT, was 1 minute 43.5, during the race on 8th September 1963. Eric Broadley drove first but not until 1.40pm, as the engine had taken some three hours to start ...

Roy Salvadori took over at 2:10pm but came into the pits on lap seven to say that: "It feels as though the rear end of the vehicle is lifting," as well as the steering feeling heavy. He then went back out and completed some 40 laps to check the fuel consumption using the Hillborn injection, never getting below a lap in 1 minute 45.4 seconds.

The next day, it was found that number five pushrod had failed, and this was replaced before the car was checked again; this time fitted with 48mm Weber carburettors, after the usual checks to suspension and brakes were carried out.

During 57 laps of this day's test, Bruce got down to 1 minute 36.7. The next day was taken up with checking out brake balance and the amount that the engine 'wound up,' which required re-working of the engine mountings. However, with a lap time almost six full seconds faster than the Aston Martin opposition, the team could feel rightfully proud of its achievements. This was summed up in Bruce McLaren's comment that "This is beginning to feel right." Bruce McLaren wrote a final report encapsulating his thoughts on the tests and their results. It was published in Roy Lunn's *GT and Sports Car Project*.

During these tests, the Lola GT Mk 6s had, particularly LGT/2, tried various air intakes and slots cut into the bodywork in an attempt to improve airflow to the carburettors, and also at the front to allow air to escape from the front fenders over the wheels.

At the start of the Lola/Ford collaboration, in July 1963, Roy Lunn had presented Eric Broadley with his own ideas about the new Ford. First on the list of changes was a clamshell-type roof, hinged at the base of the windshield and lifting at the rear of the cabin, of the type that Lunn had designed on his mid-engined Mustang I design of 1962. Eric Broadley pointed out, however, that it wouldn't be allowed on a race circuit, because, if the car rolled over, the driver could not get out.

Next, Roy Lunn redesigned the suspension, which Eric didn't like either, reasoning that this new design would upset the handling.

Then Lunn showed Eric Broadley his ideas for a body shape, based upon Lunn's previous design for the mid-engined Mustang I prototype. Eric Broadley didn't like the design, pointing out that the nose shape would allow air to flow beneath the nose and lift it off the ground at high speed. Lunn pointed out that the body had been designed by Ford's computer, with assistance from models tested in the Ford wind tunnel back in Dearborn. Eric stuck to his opinion, in which he predicted that the new Ford GT would take off and fly at a speed approaching 200mph, despite the computer back in Dearborn denying this. Eric was so convinced that this body shape was unsafe at speed, that he consulted John Wyer about it, who told him to "give Ford what they want." Unconvinced, Eric Broadley flew to Dearborn to attempt to make Ford management change their minds, but got nowhere. He was to be proved right at the Le Mans trials in April 1964 ...

As we have seen, Roy Lunn, the Advanced Concepts Manager, produced a series of confidential reports about his time overseeing what became the GT40, culminating in the book *GT and Sports*

Continued on page 87

GT AND SPORTS CAR PROJECT

TEST REPORTS CONFIDENTIAL

The following comments were submitted by Mr. Bruce McLaren after the last test at Snetterton:

GENERAL

The following remarks are in relation to the test car as it was for the two test periods on 27th. and 28th. of November; earlier test periods were too experimental and discontinuous for solid opinion to be formed.

Much of what I comment on will be duplicated on the daily test sheets, but there are one or two points which I should emphasize and in the event of further testing at Snetterton, the handling comments may help in judging the improvement in general feel and stability.

HANDLING

Steering, Springing, Shockabsorbing. The steering was much improved over earlier trials; there was no tendency to deviate from a straight line, either under heavy braking at high speed, or when using the edge of the road before or after a corner. The car had been bump steering, but is not now, with just one provision. The fitting of the low rebound rate shockabsorbers, which brought back a trace of this condition. The steering ratio still feels high (quick): only a movement of the wrists is necessary to correct a high speed slide and it is difficult to do this accurately; the high speed bend on the Mulsanne straight at Le Mans would need only pressure on the steering wheel rather than movement. Although a lower gear ratio is essential for Le Mans, the present ratio could suit Sebring. With the lower ratio in mind, I would like to see the steering wheel kept as near vertical as it is now, or nearer, making it possible to make nearly a complete turn to the right, say, with the right hand. The damper improved the general feel.

SPRINGING AND SUSPENSION

As the spring rate was increased the car used less road and lap times improved; it was not as controllable or predictable, but the cornering speed was not then limited by what felt to be considerable body roll and slide.

The increase in front spring rate did not generate any more understeer; the cornering attitude of the car became much more consistent, however. At no time has there been a strong understeer condition, and I would like to see a good deal of understeer as a possible setting for Le Mans; considerable time can be gained with an understeering car in the region of Whitehouse, Dunlop Bridge and Mulsanne corner, all very fast corners.

74

This & next two pages: Bruce McLaren's a final report encapsulating his thoughts on the tests and their results.

GT AND SPORTS CAR PROJECT

TEST REPORTS CONFIDENTIAL

Although the combination was not tried, I think that the highest rate front and rear springs with the shock-absorbers set at 1 or 2 from maximum, would be the optimum on the present basis. At Snetterton during cornering, the rear wheels could be broken loose by full throttle application in third gear; this condition is one that has been improving steadily; at Goodwood during initial testing it was possible to put the car completely sideways with the same power application.

At Snetterton on the exit from Coram Curve, the car would end up oversteering slightly on full throttle in third, and the fast left hand curve before the main straight ended up in an oversteer if full throttle was used, such that it was necessary to back off. The corner could be taken on a gradually opening throttle or 3/4 throttle entering the straight at 6,500 rpm in third.

This corner is similar to two of the airport turns at Sebring; it may not be advisable from some points of view but, more roll stiffness at the front would enable full throttle to be used on this corner.

The car is good on the hairpin turn, a gently oversteer is easy to induce on the way in, and the rear wheels spin only occasionally at peak torque in 2nd gear, (which indicated I think that the problem of the tail breaking loose at high speed is not just a question of too much power).

BRAKES

Maximum improvement has taken place in this area; the braking in itself is perfectly satisfactory now. However, the pedal travel is long (and generally ends up longer in endurance races). It is not heavy so a change in mechanical or hydraulic ratio would be acceptable. The long travel (requiring sometimes two pumps) is caused by taper wear on the pads, which at present are of rectangular section, judging by experience with the same caliper on Aston Martins, the segmental pad is not the complete answer, a greater included angle in the segment may help.

As long pedal travel can be mistaken for fade and in general is a bad thing in a race; experiments may be worthwhile.

Particularly at Sebring, much time can be made up and cars overtaken under braking, preferable to extending either Engine or Driver.

Most important is that it has become apparent that the very large front pads need considerable bedding, at least ten laps, before the initial fade period occurs; the customary two or three laps bedding may result in an initial fade during the opening stages of a race, which in close company could be disastrous. I think it is worth noting that very large capacity reservoirs are handy, once again with Sebring in mind where pad wear and brakes in general are a major factor.

75

GT AND SPORTS CAR PROJECT

TEST REPORTS CONFIDENTIAL

CARBURETION

At Sebring with sharp right angle corners, part throttle running is most important; this is the only apparent fault with the carburetion now. The engine will not run either at high or low revs on the first portion of throttle judging by the way the engine starts from cold; this is only a question of smaller idle jets, less pump travel, and synchronization. Throttle pedal rate and movement is O.K.

DRIVING POSITION

Seat and wheel relationship is near perfect for me now, but pedals and footwell are most uncomfortable; it is possible to toe and heel, only after pumping the brake pedal. Visibility is good. I would like a full harness type seat belt and a large inside door handle. The three spoke steering wheel is O.K.

76

Car Project, released only to the managers of the various departments of Ford, first of all in June, with another updated issue on 18th December 1963. In the foreword to the book, Lunn wrote that the Advanced Concepts Department was set up to:

"Develop highly featured show cars, high performance models, racing cars, and generally keep a watching brief on new developments in the automotive World. The Group will take full advantage of advanced company developments, effect liaison with outside specialists where, appropriate, and exercise vehicle builds, mainly through outside vendor sources.

"The first projects selected for the group are:

1. A racing GT car that will have the potential to compete successfully in major road races such as Sebring and Le Mans.

2. A high performance, two-seater road sports car, in the category of, and superior to, the Corvette. Design will take into account the possibility of its being developed into a low-volume model line."

I believe it is important to note Lunn's words here, as they led directly on to his desire that the forthcoming Ford GT40, which was a development of the Lola GT Mk 6, could be used on the road. On Roy Lunn's orders, Len Bailey redesigned the Mk 6's chassis to use steel, which Eric hated, as it ran completely against his belief that the car should be as light as possible. (This, despite the GT Mk 6 prototype having had a mainly steel chassis). Instead of steel, Eric Broadley had envisaged an aluminium tub, with a fibreglass roof, but Lunn was thinking that the Ford GT would also, suitably de-tuned, be sold by Ford as an exotic car for the street. Eric was only interested in seeing it as a race car. In retrospect, it seems surprising now that Ford did not build an initial run of, say 50 chassis in aluminium, to supply both Ford run cars and the initial private customer cars, given the budget that they had.

Eric Broadley never could see the point of building an unnecessarily heavy car, and this was what principally led to Lunn and Broadley falling out. Eric pointed out that, according to his contract with Ford, he had been hired as the chief design consultant but no one was listening to him. Within a few months, neither would speak with each other, and John Wyer, now working for Ford, had to relay comments from one to the other, which he found exasperating. On top of that, Eric, being an entrepreneur, couldn't deal with the red tape of the Ford Corporation and became very disillusioned with the deal that he had agreed to. By December 1963, Eric Broadley couldn't wait to get out of his contract with Ford.

Whilst these disagreements between Lunn and Broadley had been going on, the design team and engineers had been busy constructing the new Ford GT design. American engineers in the shape of Chuck Mountain and Ed Hull had been sent to England to work with the current Lola employees and, by 23rd October, the prototype bodywork was photographed in mock-up form. There are also photos, taken earlier, on the 4th October in England, of the mock up body of the forthcoming Ford GT40 mounted upon one of the Lola GT Mk 6 Chassis. The wheel fixings of the GT Mk 6 were unique, so this helps to identify what was underneath the mocked up bodywork.

On 2nd December 1963, Lee Iacocca had written an Executive Communication to Henry Ford II, AN Miller and CE Patterson, in which he told them of the acquisition of Eric Broadley and Lola's co operation in producing the new Ford GT. In this document, Iacocca not only speaks of producing what became the Ford GT40, in order to win Le Mans and the World Championship (Manufacturer's World GT Championship), but he also discusses the tests that had been carried out, and speaks of producing a Ford GT for the street, with 50-100 cars being mentioned. He saw these as being sold at around $7000 each, at a time when Jaguar E Types were less than $4000 but a Ferrari 250 GTE cost $11,500.

In his book *The Certain Sound*, John Wyer had this to say about this tumultuous period:

"In the last days of 1963, between Christmas and

Continued on page 92

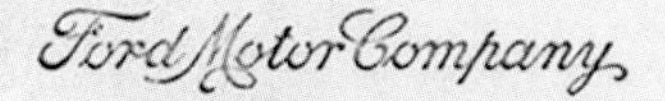

Executive Communication

December 2, 1963

To: Mr. Henry Ford II
Mr. A. R. Miller
Mr. C. H. Patterson

Subject: Status Report -- G. T. Project

On July 12, 1963, the Ford G. T. Project was approved by the Executive Office in connection with a complete review of the Division's total performance plans. You will recall that the program was approved as a replacement to the original plan to acquire Ferrari. The objectives of this project are:

I. Build a G. T. racing car capable of winning the Manufacturers' World G. T. Championship.

II. Develop and produce a "street" version of the G. T. racing car for limited domestic sale.

III. Incorporate selected G. T. design concepts in future Ford Division production vehicles.

The following specific progress has been made in accomplishing these objectives:

Objective I

Vehicle Development Contractor

Following a study of various special car builders on the Continent, the decision was made to contract Lola Cars, Ltd. of England to work under the direction of Ford engineers in the development of the initial prototypes. Lola Cars, Ltd. was selected principally because of work they were already doing in the G. T. field which closely paralleled the initial design concepts of the proposed Ford vehicle.

Personnel

Mr. Eric Broadley, Managing Director of Lola Cars, Ltd., is an internationally-known racing car builder and has had substantial experience in the designing and development of Formula Junior racing cars, as well as the original Lola G. T. mentioned above.

Mr. John Wyer, former Managing Director of Aston Martin of England and one of the best known racing team managers in the world, has been contracted to serve as Resident Manager - Special Vehicles Activity supervising all aspects of the G. T. Project, as well as other special vehicle activities in England.

Lee Iacocca's Executive Communication to Henry Ford II, AN Miller and CE Patterson. (Courtesy Ford Archives)

- 2 -

Personnel (continued)

Mr. Roy Lunn, project engineer in charge of the development of the original Mustang car, was appointed Advanced Concepts Manager for the Special Vehicles Activity with over-all engineering responsibility for the Ford G. T. vehicle.

Four Ford engineers were selected to work under Mr. Lunn's direction on all design aspects of the vehicle.

Facilities

A lease was secured on a new 5600 sq. ft. facility in a highly-skilled labor area approximately 20 miles west of London. This facility was occupied the first week in November by the members of Lola Cars, Ltd., as well as the staff of Ford personnel working on the project.

Vehicle Status

A prototype with an "Indy" push rod engine, a Ford-designed chassis but with a Lola-designed body has been under development and tested during the past three months. This vehicle set a new unofficial G. T. lap record at the Goodwood, England Track and approached the record at the Monza, Italy Track. Further refinement, particularly in Ford suspension, is now underway. A fiberglass full-sized model of the final Ford G. T. Styling concept is now in England and procurement of prototype bodies, based on this model, has begun. The vehicle will undergo further tests with the objective of final testing at Sebring in late February prior to entry in the Sebring 12-Hour Endurance Race on March 21.

It is planned that immediate steps will be taken following Sebring to produce a quantity sufficient to be qualified to meet FIA requirements for production G. T. competition (50 to 100 units). Prior to this time, a team of two of these cars will be entered in additional G. T. prototype races, such as, the Le Mans 24-Hour Endurance Race in June.

Nomenclature

In view of the substantial domestic, as well as world-wide, benefits anticipated in strengthening the Company's performance image, we propose, with your approval, to race this car as a factory entry under the name FORD GT.

- 3 -

Objective II

The domestic market for a high performance two-seater G. T. automobile has, up to this time, been captured principally by foreign manufacturers, i.e., Jaguar, Ferrari, Aston Martin. On the basis that substantial added performance image can accrue to the domestic manufacturer who builds an entry for this market, it is planned to sell, in this country, a "street" version of the Ford G. T. racing car. These vehicles, priced in the $7,000 range, will be marketed through the selective franchising of a limited number of Ford and Lincoln-Mercury dealers.

Studies will be launched immediately after the first of the year to establish the most prudent plan to produce the body, chassis and engine for these vehicles. Styling is now working on the final mock-up of the "street" version. The vehicle would be introduced as a 1965 1/2 model as originally planned on the attached timing chart.

Objective III

At the time the original concept of the present G. T. Project was developed, one of the most important advantages was to utilize this vehicle to develop future design concepts for Ford production vehicles. This was felt to be especially true in the area of suspension design. With the introduction of the T-5 next year, the first of what is hoped will be many G. T. adaptations is planned.

Steps have already been taken to design an independent rear suspension for the T-5 based directly on the Ford G. T. rear suspension. This vehicle will provide us an outstanding entry for domestic sports car racing and rallying and, on the basis of meeting initial retail price objectives, should generate substantial Company profit beginning in mid-1964. Additional design concepts of the Ford G. T. car will be considered in connection with future model plans, particularly for the Thunderbird. In this connection, adjustable pedals, unique door contours and fixed seating arrangement are all items being given consideration for production vehicles of the future.

Other Plans

Initial discussions have been held with Mr. Lou Reynolds of Reynolds Metals, Inc., who has evidenced a great deal of interest in providing complete aluminum bodies for the initial vehicles with Reynolds' bearing the cost as a merchandising and promotion venture. The benefits of such a vehicle being

- 4 -

Other Plans (continued)

fabricated in aluminum are such that we are actively engaged in finalizing some type of an arrangement with Reynolds, if possible.

For your further information, we are attaching details on this project together with photographs of the car during its testing and development. We will continue to keep you apprised of the progress of this important project in the future.

Lee.

L. A. Iacocca

The GT40, about to be shipped by air to its press launch in early 1964. From the left, John Wyer, Eric Broadley and Roy Lunn. Only Roy Lunn appears to be smiling ... (Courtesy the late Graham Robson)

New Year, I called Frank Zimmerman to tell him that the rift between Lunn and Broadley had reached a stage where it was endangering the project, and that to resolve it would require a radical redefinition of responsibilities. Since that was clearly beyond my authority, I suggested that, as a matter of urgency, he came to England."

Frank Zimmerman duly flew to England, arriving on 6th January 1964, and stayed for two days. Together, he and John Wyer visited AC Cars and Lotus (presumably to discuss the AC Cobra and the Lotus Indycar projects – Author), and he "spent a short time talking individually to Roy and Eric" (John Wyer, *The Certain Sound*). When he went to the airport to catch his flight back to America, he left the resolution of the Lunn/Broadley impasse to John Wyer.

On 13th January 1964, Roy Lunn wrote a report to Frank Zimmerman. Besides admitting that the build of the new Ford was "well behind schedule," Lunn also enclosed an organization chart, showing Eric Broadley reporting to him, which had not been part of Frank Zimmerman's original definition of responsibilities, laid out the previous September, and, according to John Wyer "Was hardly calculated, in view of their strained relationship, to promote the essential goodwill."

John Wyer went on to record that he then wrote to Frank Zimmerman at the end of January, saying that Eric Broadley, who had received a copy of Roy Lunn's report, would in no circumstances sign the contract as drafted, which was intended to be for two years. However, John Wyer wrote, Eric Broadley was "prepared to sign a contract, to end in June 1964, which would give him undisputed responsibility to Ford for the technical direction of the project. He also was happy to accept administrative and financial direction, but insisted upon technical freedom within the broad lines of engineering policy." Eric Broadley acknowledged that this would see Ford through the building of the prototype and the first three race cars.

Furthermore, according to John Wyer, Eric Broadley stipulated that Ford, "if they wished to remain at Lola's factory at Slough, should be responsible for the removal expenses of Lola Cars to a new location; that he should be adequately compensated for the loss of the second year of the contract; and that, after 30th June 1964, no restriction should be placed upon his activities and that he should be free to develop GT cars, with or without Ford." John Wyer concluded this letter to Ford by saying: "In the circumstances which have unfortunately arisen he [Eric Broadley], does not feel able to work happily or to do justice to himself on the project and one can only respect his views."

John Wyer and Eric Broadley flew to Ford headquarters in Dearborn, Michigan, on 10th February 1964. On the night of his arrival, John Wyer had dinner with Don Frey and Frank Zimmerman, and suggested that the arrangement with Eric Broadley should be terminated, and that Ford should set up "our own organization with a proper corporate existence. This was agreed and it was left to me and to Ray Geddes to negotiate terms with Eric."

Upon learning, via John Wyer, of how dissatisfied with Ford Eric Broadley had become, Henry Ford II terminated the existing contract and instructed his accountants to pay Eric Broadley in full in the autumn of 1964, and furthermore to give him the new factory at 826 Yeovil Road, on the Slough Trading Estate, that Lola had moved into in order to design and build the new Ford GT car. This was next to the new Ford Advanced Vehicles factory that would be set up to be headed by John Wyer, and which was to build and sell the new GT40. Wyer had signed a three year contract with the Ford Motor Company to look after the cars to be raced in Europe, and then to sell and service the road-going version intended to follow this. John Wyer again:

"These negotiations duly took place, with Eric Broadley receiving the whole of the new factory at 826 Yeovil Road in Slough (Ford had been using half of it to design and build up their new car). Furthermore, this new contract, which would run from 1st July 1963, meant that Lola Cars Ltd would be paid, including management charge and profit guarantee, for a full year. Finally, we agreed to compensation for the second year of the aborted contract. Let no one say that Eric Broadley had a bad deal from Ford, and, to his credit, Eric himself has never suggested it."

Eric Broadley signed the revised contract. He then swiftly got down to designing his next project, the Lola T70, when his contract with Ford expired in July 1964.

John Horsman, who was taken on at JW Automotive (an amalgam of John Wyer and John Willment's, the two founders, initials), as an engineer, started at Ford Advanced Vehicles, the factory used to build the customer GT40s, and situated next to the new Lola factory, in early July 1964. John Horsman wrote in his book *Racing in the rain*:

" ... as soon as I had graduated from the course in

39

General Specification and Description - Ford GT

Configuration: Midship engined, two seater employing the Ford 4.2 litre, 350 hp Indianapolis engine.

Dimensions: Wheelbase 95 in. - Track 54 in. - Length 159 in. - Maximum Width 70 in. - Height 40.5 in. - Ground Clearance 4.8 in.

Weights: Complete car, less driver and fuel, 1820 lbs. Front to rear weight distribution 43-57%

Components Specifications & Suppliers:

Component	Supplier
Main body structure -- light gauge monocoque	Abbey Panels Ltd. Coventry, England
Body ends, doors, dash -- thin fibreglass fabrications	Specialised Mouldings Ltd. Upper Norwood, England
Instruments -- tachometer, speedometer, oil pressure, oil temperature, water temperature	Smiths Motor Accessories Ltd., Cricklewood, England.
Body electrical equipment -- 12 volt wiring, switch gear, 2-speed wipers, electric washer, horn, ampmeter, turn indicators	Joseph Lucas Ltd., Birmingham, England
Rectangular headlamps using iodine vapour bulbs, 2 pass lights	Cibie, Paris, France.
Carburettors -- twin choke down-draft 4? mm units	Edoardo Weber s.p.a. Bologna, Italy
Exhaust pipes -- individual pipes culminating in two tuning chambers and outlets	V. W. Derrington Ltd. Kingston on Thames, England.
Plugs	Autolite Division Ford Motor Company Wixom, Michigan, U.S.A.
Suspension -- independent link types using welded tubular arms and low friction bearings	Southwest (Bearings) Products, Monrovia, California, U.S.A.
	Shaffer (Bearings) Ltd., Division of Chain Belt Corp. Dollars Grove, Illinois, USA
Shock absorbers -- adjustable telescopic	Armstrong Patents Co. Ltd. York, England.
Road springs -- coil located around shock absorber	Tempered Spring Ltd., Sheffield, England.

Above & opposite: A list of the suppliers Ford engaged in the GT40 project; note how many are based in England. (Courtesy Ford Archives)

London [business management at the London School of Economics – Author], on June 30th of 1964, I arrived at the Yeovil Road premises of what was still Lola Cars, on the Slough Trading Estate."

Then, speaking of what had gone on before he arrived, Horsman wrote:

"Unfortunately there was *no clear definition of*

- 2 -

Components Specifications & Suppliers:		
	Wheel bearings — 2 tapered roller units per hub	Timken Roller Bearing Co. International
	Steering -- rack and pinion	Cam Gears Ltd., Luton, England
	Transaxle -- 4-speed transmission combined with axle unit in an aluminium housing	Gear Speed Developments, spa, Modena, Italy.
	Clutch -- 3 plate $7\frac{1}{2}$" diameter hydraulically operated	Borg & Beck Co. Ltd., Leamington, England
	Brakes -- $11\frac{1}{2}$" discs with aluminium calipers using separate systems front and rear	Girling Ltd., Birmingham, England.
	Brake Linings — DS11	Ferodo Ltd. Chapel-en-le-Frith, England
	Wheels -- 15" aluminium rims on wire construction with rudge centres	Costruzioni Meccaniche, s.p.a., Milan, Italy
	Tyres -- 7.25 x 15 rear 5.50 x 5.15 front	Dunlop Rubber Co. Birmingham, England
	Cooling system — front radiator combined with oil cooling unit	Serck Radiator Services Ltd. London, England.
	Fuel system -- plastic bags housed in body slide members	Goodyear Tyre & Rubber Co. Akron, Ohio, U.S.A.
	Fuel pumps — bank of three electric	Bendix Eclipse Machine Corp. Elmira, New York, U.S.A.
	Battery — 12 volt 52 amp source	Autolite Division Ford Motor Company, Wixom, Michigan, U.S.A.
	Driveshafts — tubular with Cardan joint at outer end	B.R.D. Co. Ltd., Aldridge, England.
	Inboard couplings — rubber rotaflex units	Metalastic Ltd., Leicester, England.

responsibilities from the outset [Author's emphasis], resulting in order and counter-order between Broadley and Ford engineer Roy Lunn. When I arrived at the Yeovil Road workshop, neither of these two great designers was speaking to the other, instead communicating through middleman John

The first GT40, X40-101, when first built. (Courtesy Ford Press)

Wyer. Wyer was nominally in charge of the (Ford GT40) operation, but even that changed from one Ford management meeting to the next, depending on internal politics.

"At the time of my arrival the operation was still being run under the Lola name and bank account, but Ford Advanced Vehicles (FAV), with Walter Hayes and Sir Leonard Crossland from Ford of Britain, and Wyer as directors, was finally formed on 1st July 1964."

However, before the final falling out between Eric Broadley and Roy Lunn occurred, Lola had been joined in the autumn of 1963 by Len Bailey, Ron Martin and Chuck Mountain, all from Ford of America. Broadley and Bailey were charged with design of the chassis, Ron Martin was the body engineer, and Phil Remington came over from the Shelby organization to oversee the engine installation.

Len Bailey was another ex-English engineer, working at the time in the car research group at Ford in Dearborn. Prior to this, he had belonged to the 750cc Motor Club in England, as had Eric Broadley.

About Len Bailey, John Wyer had this to say:

"Between September 1963, when we effectively started work in England, and the end of March 1964, we completed the design of the Ford GT40 and the build of the first two cars. It was, in the circumstances, a remarkable achievement, for which credit must go to Eric Broadley himself and, in particular, to Len Bailey, who, with his very small staff, was so largely responsible for the execution of the GT40 design. Len, a design engineer of exceptional ability and talent, refused to allow himself to be distracted by the political in-fighting and never relaxed his almost super-human efforts to get the job done."

After his collaboration with Lola and Ford, Len Bailey and his wife stayed in England, and he later

The first GT40 being prepared for shipping. Behind it, in the corner, can be seen the Prototype Lola GT Mk 6, LGT/P. (Courtesy Allen Grant)

redesigned the GT40 into the Mirage, for the JWA racing team, before also designing a BRM V12-engined prototype, the M2, and then moving on to other engineering projects.

Chuck Mountain had been working on vehicle dynamics and suspension design for the Ford Scientific Lab and Research Group. Chuck was in charge of re-imaging the GT Mk 6 as the Ford GT40. Ed Hull, another Ford engineer, helped with suspension design. Ron Martin designed and built the first Ford GT bodies, using fibreglass.

Incidentally, after the Ford GT40 had been announced to the press in April 1964, the Ford Motor Company of America later (probably around 1966), published a poster showing the development of that car, which completely missed out the Lola GT Mk 6, and instead showed the GT40 being descended directly from the Roy Lunn-designed Ford Mustang I project ...

Laurie Bray said:

"The Ford GT40 was ... heavy. It did have a very nice engine, though, that 4.2-litre V8 was excellent. That went into all the early GT40s before Ford employed the Cobra's 4.7-litre engine.

Upon the announcement of the new Ford GT40, Carroll Shelby was enlisted to give rides in the new car to various Ford Executives. (Courtesy Ford Archives)

"After the GT40/Lola contract was terminated in July 1964, Ford did ask Terry, Andy and I to go and work for them but we declined – Eric was a great innovator, even if he sometimes didn't understand how to make something work! That cable-operated gear change on the GT Mk 6 is a case in point, it never worked properly."

Lola Cars Ltd had by then moved to a new factory at 826 Yeovil Road on the Slough Trading Estate, as the old premises in Bromley were completely inadequate to house the number of engineers that Ford insisted were needed to help Eric Broadley design its new Ford GT. Here Lola personnel first of all built their last Mark 5As, later to be called the Mark 53, for the Midland Racing Partnership of which Richard Attwood and David Hobbs were members. Although the new factory was in the name of Lola Cars Ltd, it also encompassed what would become Ford Advanced Vehicles in one half of the building. This is where the first GT40s were built, all by Lola personnel.

By late August 1963, the first sheet metal was cut at Abbey Panels. The first GT40 build was started, and ten days later the second car build was commenced. The new Ford GT40 rolling chassis were completed by October. The engineers then had to wait until the bodywork was delivered, sometime in December. Maryland University had provided its wind tunnel in which to mount the 3/8 scale model, but it couldn't measure speeds over 150mph, so quite how the car would behave at its 200mph top speed was, at the time, unknown. After the Le Mans trials in 1964, during which the GT40s became unstable at speed, a lower nose was designed and a tail spoiler was fitted. The Ford GT40 then became drivable at the high speeds of which it was capable.

Eric Broadley, having now completed his contract with Ford in July 1964 (the new Ford being designated the GT40), returned to his own company in Yeovil Road, Slough and commenced designing what became the Lola T70 sports-racing car, in September 1964.

Chapter 6

Allen and the Lola GT

To understand the reasons behind Allen Grant's long ownership of the prototype Lola GT Mk 6, you have to understand what drives him. Most of that came in his youth, when he set out on the path to becoming a professional race driver.

Allen met a certain George Lucas in 1959, who,

Santa Barbara in 1963: Allen Grant waves at the camera whilst George Lucas, his crew chief, rides with him on a lap of honour after winning yet another race, this one at Santa Barbara. (Courtesy Allen Grant)

Allen Grant at Riverside in 1964. (Courtesy Allen Grant)

Le Mans 24 Hours, 1965, prior to the start. From left-to-right: Carroll Shelby, Alan Mann, Allen Grant, Jo Schlesser. (Courtesy Allen Grant)

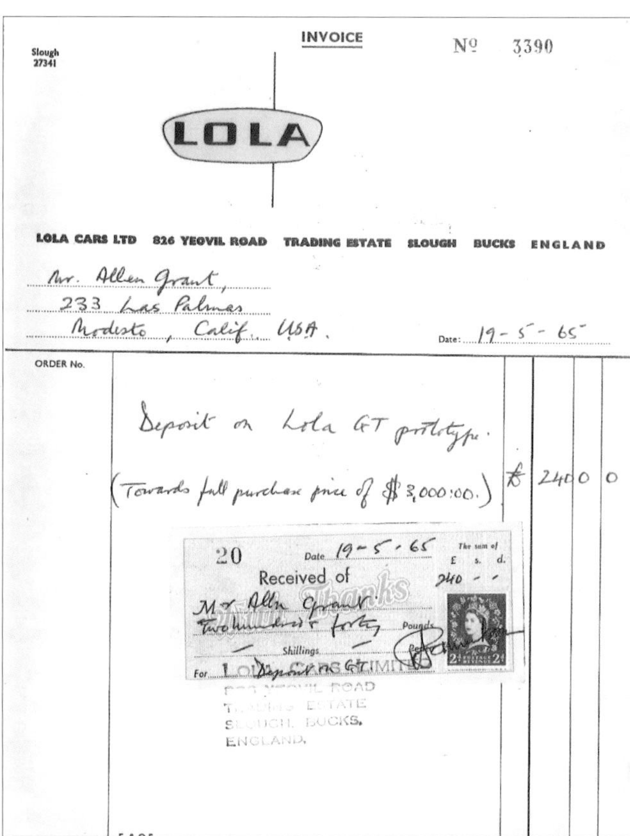

Slough 27341

INVOICE

Nº 3390

LOLA

LOLA CARS LTD 826 YEOVIL ROAD TRADING ESTATE SLOUGH BUCKS ENGLAND

Mr. Allen Grant,
233 Las Palmas
Modesto, Calif. USA.

Date: 19-5-65

ORDER No.

Deposit on Lola GT prototype.

(Towards full purchase price of $3,000:00.) £ 240 0 0

20 Date 19-5-65 The sum of £ s. d.
Received of 240 - -
Mr Allen Grant
Two hundred & forty Pounds
— Shillings — Pence
For Deposit on GT
LOLA CARS LIMITED
826 YEOVIL ROAD
TRADING ESTATE
SLOUGH, BUCKS.
ENGLAND.

The invoice, still in the possession of Allen Grant from when he bought LGT/P in May 1965. (Courtesy Allen Grant)

when he made *American Graffiti*, relied heavily upon Grant's experiences while the pair were street racing in Grant's home town of Modesto, California. In fact, Lucas modeled the part played by Paul Le Mat.

After cruising 10th Street in Modesto in a 1950 Pontiac, Allen tried driving a friend's MGA, and that turned him on to European sports cars. In September 1959, he bought a new Austin Healey 3000 and started competing in autocross events. He won straight away, but then he rapidly discovered that the car to have was actually an AC Bristol. Allen swiftly exchanged the Healey for an AC Bristol and started racing more with the SCCA. In his first year of racing, he won 12 races from 14 starts.

Having read of Carroll Shelby's decision to re-engine the AC Ace with a Ford V8 engine, Allen, in January 1963, visited Shelby at his premises in Venice, California. Although he was turned down to be a Shelby racer, Phil Remington offered him a job as a welder, which he accepted.

Whilst at Shelby's, Allen sold a competition Cobra to Coventry Motors of Walnut Creek, with the proviso that he had to be the driver when it raced. George Lucas was his crew chief and designed the livery. Right away, the yellow Cobra won "out of the box." The highlight of the year was the *Los Angeles Times* Grand Prix where, on the first lap, Allen was tapped into a spin when lying second, by Shelby

Above & overleaf: Later years, Allen Grant. (Courtesy Allen Grant)

driver Bob Bondurant. Recovering dead last, Allen drove the doors off the Cobra, to place second overall in the one-hour race. The *Los Angeles Times* wrote that Grant "went through the field like a constipated bull(!)" That earned him a ride with the factory team for 1965.

But then ... Allen was drafted into the US Army and didn't get back to Shelby's factory until May 1964, by which time Shelby had all his drivers hired for the '64 season. Carroll wanted Allen back in an administrative position, but that wasn't in Allen's plans, and so he accepted a drive in Bill Thomas' 'Cheetah,' a car with the engine mounted so far back in the chassis, it gave it some ill-handling traits.

Allen was back in a Cobra for the '65 season, although this time it wasn't the open roadster Cobra but the Daytona Coupé version, designed more for top speed than the blunt, open bodywork of the roadster. Paired with his old teacher, Ed Leslie, Allen placed third in the Daytona Continental, then finished in the 12 Hours of Sebring, despite clashing with an errant Volvo.

For his first race in Europe, Allen was paired with old adversary Bob Bondurant in an Alan Mann-entered Daytona Cobra, chassis number CSX2300, in the Monza 1000km, and they won the GT class, handily beating the previously all conquering Ferrari GTOs. Allen Grant was back in a Radford Garages'

open Cobra for the next race, the Tourist Trophy at Oulton Park. "The car was a dog," remembers Allen, but he still finished sixth overall.

It was during May of 1965 that Allen visited Ford Advanced Vehicles in Slough, whilst working on the Cobra Daytona Coupé at Shelby's behest, during his time in England. Allen had gone to Ford Advanced Vehicles together with Charlie Agapiou, and Gordon Chance, two other mechanics who were all then with Shelby's team. Allen wanted to make sure that he prepared the Cobra Daytona Coupé that was to drive in World Championship events that year.

Whilst next door at the Lola factory, he spotted the Lola GT Mk 6 prototype, LGT/P, in a corner of the workshop, and negotiated a price of $3000 for it, minus gearbox, with Rob Rushbrook, Lola's workshop manager. He has owned it ever since.

Allen's last professional race was the 1965 Le Mans 24 Hours, where, paired with Jo Schlesser, the pair retired early on after the clutch failed. Allen went back to complete his education in America and to become a builder and property developer. He had one more try at motor racing, putting together a two-car Formula Ford team for Jimmy Vasser and Ken Murillo in 1989.

Chapter 7

Afterwards

This chapter is dedicated mainly to describing the histories of each of the three GT Mk 6 Lolas that were built. However, before that, there is the story of 'what happened' with the Lola Mk 6 design after Lola and Ford had parted ways in 1964.

The story of the Ford GT40, directly derived from the Lola GT Mk 6, is well known. Although it initially failed, in the 1964 and 1965 races, in 1966 Ford finally won Le Mans; GT40s taking the first three places, with their 427in^3 engined Mk 2s, although in a controversial finish. They won again with a further works effort with the Mk 4 in 1967 and then Ford, as a factory entry, retired.

John Wyer's company, JW Automotive, then took over the Ford racing programme for Europe and, in 1968 and '69, won Le Mans with a GT40 using a small block-engined GT40, the same car winning on both occasions. JW Automotive then switched to running Porsche's factory effort with the Typ 917, and, though it did well in the World Championship races, JWA failed at winning the Le Mans 24 Hours for Porsche, although other private teams did.

Roy Lunn carried on as a designer/manager at Ford. In 1962, he had designed, for Ford, his 'Mustang 1,' a mid-engined creation that led on to the production, front-engined, Mustang of 1964, which proved to be a huge sales success for Ford. In 1966, he produced the *SAE Papers* (Society of Automotive Engineers), describing the success of the GT40 and basically airbrushed the Lola part of it out of the story, showing the GT40 being descended from his 1962 'Mustang I.' His encounters with Eric Broadley obviously did not leave him with good memories …

Allen Grant standing beside his Lola GT Mk 6 upon its arrival in Los Angeles, around 1965. (Courtesy Allen Grant)

Eric Broadley went back to designing race cars for his own company, Lola Cars Ltd, and was consistently successful in this over the next 30 years, his next car after the Mk 6 being the hugely successful Lola T70, which he commenced in September 1964. This was a car that was successful from the start, and is today loved and revered by all who have driven it. Amongst the T70s initial successes was winning the 1966 Can Am Championship with John Surtees.

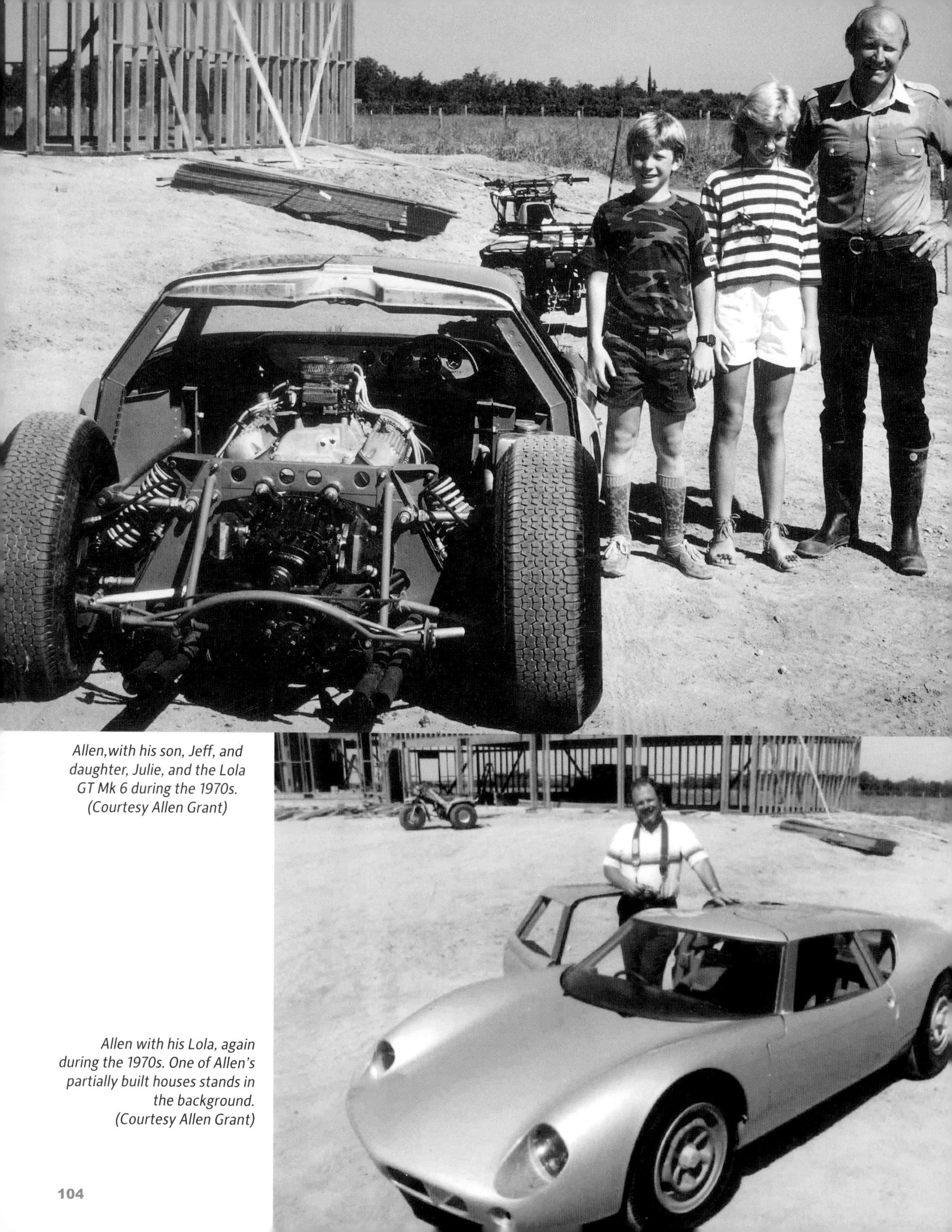

Allen,with his son, Jeff, and daughter, Julie, and the Lola GT Mk 6 during the 1970s. (Courtesy Allen Grant)

Allen with his Lola, again during the 1970s. One of Allen's partially built houses stands in the background. (Courtesy Allen Grant)

The interior of a Lola GT Mk 6 in 2012. (Author's collection)

Incidentally, following the GT Mk 6, this should chronologically have been the Lola Mk 61 but the Elva Mk 61 was already so named, and Lola changed its type numbering by adding a '0' after the 7, thus ushering in the new type numbering system that allowed for different modifications to each subsequent Type to be so delineated, such as the Type 160, 162, 164, 165, etc, all being variants of the original Type 160 Can Am car. Lola also posted many successes at the Indianapolis 500, amongst myriad other races. Now back to the Lola GT Mk 6 ...

All three examples built have survived, as befits a very notable type. The first is L (for Lola) GT/P. This was the prototype, shown at the Racing Car Show in London in January 1963. It raced at Silverstone with Tony Maggs driving on 11th May, and two weeks later it raced in the Nürburgring 1000km with Maggs and Bob Olthoff.

After this LGT/P was sold to the Ford Motor Co of Dearborn, USA, but stayed in England, taking part in various tests (detailed elsewhere in the book), with Eric Broadley, Bruce McLaren and Roy Salvadori driving.

This prototype GT Mk 6 formed part of the separation agreement between Lola and Ford, and was returned to the Lola Company in 1964. Allen Grant of Shelby American saw the car at the Lola factory in Yeovil Road, Slough when he was working next door at Ford Advanced Vehicles. He purchased the Lola in May 1965 and has owned it ever since. Allen had the GT Mk 6 renovated to as-new specification in 2009.

Such an evocative shape. The beautifully proportioned front of the Lola GT in 2012, still looking breathtakingly modern. (Author's collection)

Allen Grant's beautifully restored Lola GT Mk 6, chassis number LGT/P. (Courtesy Evan Klein)

In 2017, Allen Grant drove his freshly restored Lola GT Mk 6 at the Goodwood Festival of Speed. (Courtesy Steve Havelock)

The crowds at Goodwood's Festival of Speed loved Allen Grant's Lola GT Mk 6. (Courtesy Steve Havelock)

Lola GT Mk 6 Coupé, Chassis # **LGT/P**
Prototype, built in December 1962/January 1963.

1963:
26/01: London Motor Racing Show exhibit. Silver.
11/05: Silverstone, Daily Express Trophy: T Maggs, #48; 9th OA.
25/05: Nürburgring 1000km: T Maggs/B Olthoff; #115; DNF. Silver w/green stripe. Rear air ducts and mirrors added after practice.

??/08: Sold to Ford Motor Co.
08/10: Brands Hatch, Goodwood, Monza, Snetterton: Tests with Bruce McLaren.

1964:
07/64: Returned to Lola Cars Ltd.

1965:
19/05: Sold to present owner (Allen Grant), minus engine and gearbox, for $3000. He went to Specialized Mouldings to find the original tail.
??/07: Shipped to Shelby American in Los Angeles.

1981:
04/10: Displayed at the Modesto Concours d'Elegance, California.

1998:
07/09: Displayed at the All-British Field Meet, Portland, Oregon.

2008:
/01: Renovation begun.

2009:
Renovation 90 per cent completed.

2016:
Renovation complete.

2017:
Goodwood. Festival of Speed, driven by Allen Grant.

Lola Mark 6 GT Coupé, Chassis # LGT/1
1963:
Dark green with silver longitudinal stripe.
Street registered: 0142 KE.
Driven on road to Le Mans by Eric Broadley.
12-13/06: Le Mans 24 Hours: R Attwood/D Hobbs, #6; DNF (acc).
Repaired.
05/08?: Guards Trophy, Brands Hatch: A Pabst, # 3; DNF (engine). (See LGT/P).
Sold to the Ford Motor Co, USA.
08/10: Used as a test mule for the Ford GT40 programme at Goodwood, Monza and Snetterton.
Then to Alf Francis, (Colotti Gearboxes), in Italy.
Then sold to Serenissima, where Francis then re-badged and re-engined it with a Serenissima engine and claimed as one of their own, whilst in Modena.
To a Ferrari dealership in Oklahoma City, OK, USA.
Sold to James Whitmer, Oklahoma City, OK.
Afterwards, it received a Ford 289 engine and a GT40 tail.
Sold to Jerry Benson, Vancouver, BC, Canada.
Raced at Westwood Racetrack, crashed.
Put in a container for many years.

2000:
Sold to Jerry Bensinger, Youngstown, OH.
Sold to Paul Haywood-Halfpenny and Partner, UK.

2001:
Rebuilt by Barry Hodson Motorsports, UK.
23/05:Raced at Le Mans Legends by Paul Haywood-Halfpenny, #6; DNF.
Raced at Snetterton by Tony Dron and Paul Haywood-Halfpenny, #17; 1st.
08/07: Goodwood Festival of Speed: P Haywood-Halfpenny, #210.

2002:
12/07: Auctioned by Bonhams at Goodwood, sold to Hideako Suzuki, Japan.

2004:
24/07: Le Mans Classic: H Suzuki/M Kotaro, #62.

2020:
Still with Hideaki Suzuki, Japan.

Lola Mark 6 GT Coupé, Chassis # LGT/2
Finally, to the third and last Lola GT Mk 6 built, LGT/2. This car was sold from Lola (much to Ford's displeasure, but before the Lola/Ford deal took place), to John Mecom, a wealthy Texan oilman,

Montecito
2021
HAMMER TIME
353

Success! Allen Grant with his 1963 prototype Lola GT Mk 6 after winning the 2021 edition of the Montecito Concours. (Courtesy Allen Grant)

who became Lola's distributor in the USA for the next few years, before Carl Haas took on the job in August 1967.

Augie Pabst raced the car in 1963 at Nassau where a Traco-tuned Chevrolet 327ci engine was used. In December, again driven by Pabst, it won two races at Nassau during the 'Speed Weeks.'

In October 1964, after a string of DNFs, LGT/2 was crashed at Riverside. The car was sold and repaired, going to two more American owners before being bought and restored by Gordon Gemball in 1987.

Sold to Peter Kaus in 1988, and spent many years in the Rosso Bianco Collection in Aschaffenburg, Germany. In 2006, the GT Mk 6 was bought by well-known collector and vintage racer Bruce McCaw. It has since been sold to the Bardinon Collection in France.

1963:
Sold to John Mecom. Ford engined.
Traco Chevrolet 327ci V8.
Painted metallic blue with white longitudinal stripe.
??/08: Tested at Brands Hatch.
05/08: Guards Trophy, Brands Hatch: A Pabst, #3; DNF. Engine.
Ford engine then replaced with Chevrolet 327ci V8 engine.
01/12: Nassau GT five-lap race: A Pabst, # 00; 1st.
01/12: Tourist Trophy, Nassau: A Pabst, # 00; 1st.

1964:
21/03: Sebring 12 Hours: W Hansgen/A Pabst, # 20; DNF (engine).
06/06: Player's 200, Mosport: A Pabst; #00; DNS (engine blew in practice).
06/06: June Sprints, Road America: A Pabst, # 2; DNF (overheating engine).
03/08: Guards Trophy, Brands Hatch: A Pabst, # 8; 11th.
??/08: To Lola Cars Ltd, Slough for wider wheels and bigger brakes.

Four views of Lola GT Mk 6 **LGT/1** *when owned by J Whitmer in the 1970s/80s. It has been 'tricked out' to very much resemble a Ford GT40.*

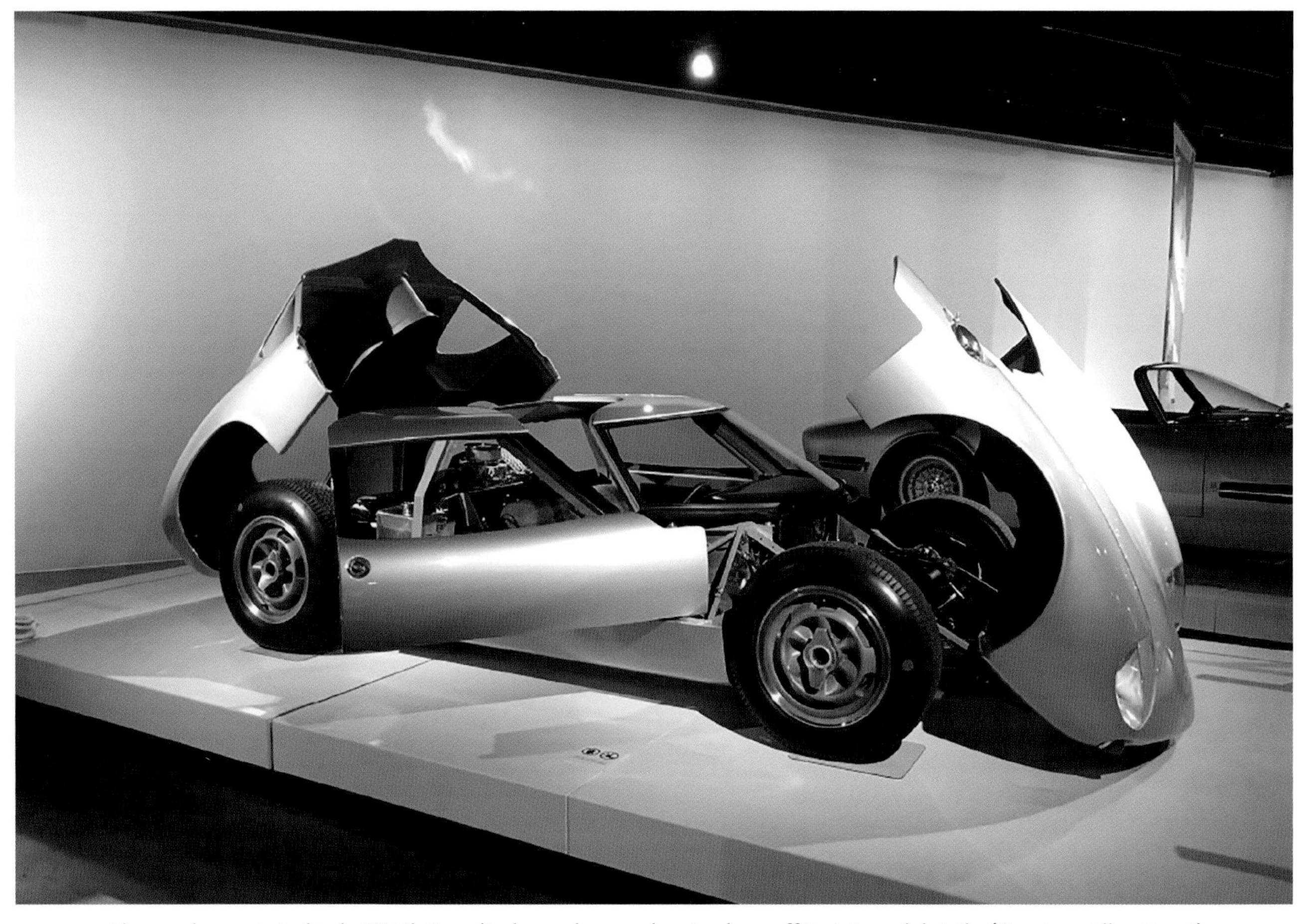

The newly renovated Lola GT Mk 6 on display and opened up to show off its internal details. (Courtesy Allen Grant)

13/09: Road America 500: A Pabst/W Hansgen, #1; DNF (engine).
11/10: LA Times GP Riverside: A Pabst, # 24; DNF (acc)
Sold to Dave Dunlop, Austin, TX.
Sold to Larry Crosson, Sacramento, CA.

1987:
Sold to and Restored by Gordon Gemball, Sacramento, CA.
Sold to Wayne Lyndon, Sacramento, CA.
Sold to Chuck Haines, MS.

1988:
Sold to the Rosso Bianco Collection, Germany.
Rebuilt by Tom Fredericks, Brooks, CA.
Displayed at the Rosso Bianco Collection, Germany.

2006:
01/09: Sold to Bruce McCaw, CA.

2012:
12-15/02: Retromobile show, Paris.
04/08: Sold to Dubai.

2016:
30/08: Monterey auction: $1.2m bid, not sold at auction, sold afterwards to France.

2022: In the Bardinon Collection, France.

Chapter 8

Renovation

In the spring of 2008, Allen Grant, the longtime owner of the prototype Lola GT Mk 6, decided that it was time the car was restored. Fortunately, Allen owned his own shop in Washington State, and decided to participate in the restoration himself, using his own shop to do it in.

To assist, he brought in Rob Senekal; probably the pre-eminent authority on the construction of Ford GT40s, which were the descendants of the Lola. Rob, or 'Robbie' as he is usually known, was from Johannesburg in South Africa, and served an apprenticeship as a tool and jig maker whilst attending technical college. Along the way, Robbie ran his own pattern shop and metal pressing business, and he worked with the Superformance agent for Ohio. Via a 15-year spell as a chemical engineer at the University of Capetown, in the role of chief technical officer, he joined CAV as its chief engineer, and, in March of 2002, became general manager/chief engineer on the Hi-Tech Automotive/ Superformance GT40 Mk II project.

John Hill made up the other partner of the trio. John was a friend of Allen's, and had restored American cars, and then turned his hand to restoring rotary-engined Mazdas.

Although the renovation had started in 2008, it lapsed whilst Allen moved from Washington to Palm Springs, whereupon it started again in Allen's new shop there in 2016. The Lola, in dismantled form, had been trucked overland from the North West to the South West. The process was called a renovation, as against a restoration, as there was no crash damage, no rust: everything simply required taking apart, cleaning, and then careful re-assembling.

Renovated to its original specification when it was first seen at the London Racing Car Show in January 1963, and completed just in time to take part in the 50th Anniversary of the Ford GT40, the Lola made its debut at the Quail in Monterey in 2016, and won its class handsomely. David Lillywhite, editor of *Octane*, gave it the 'Editors Award,' and Allen Grant drove it for the very first time onto the winner's rostrum.

The following photos and captions are a snapshot of the work that was carried out to bring the Lola GT Mk 6 back to life.

The chassis before painting. (Courtesy Marc Willard)

Wishbones (A-arms), coil spring enclosed dampers, brake calipers and discs and the dash can all be seen on the bench. (Courtesy Marc Willard)

Left & above: It takes a lot of parts to build a car! (Courtesy Marc Willard)

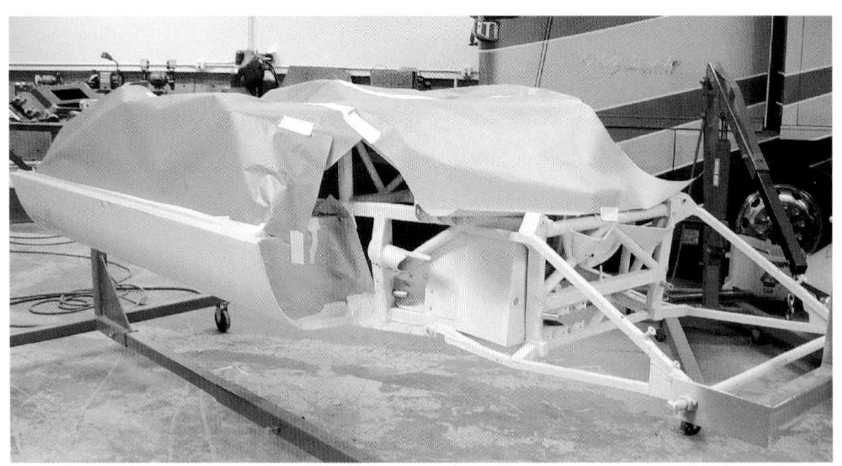

The renovated chassis in primer, awaiting painting. (Courtesy Marc Willard)

The chassis before painting. (Courtesy Marc Willard)

The chassis after painting. (Courtesy Marc Willard)

The rear bulkhead of the renovated chassis before painting. (Courtesy Marc Willard)

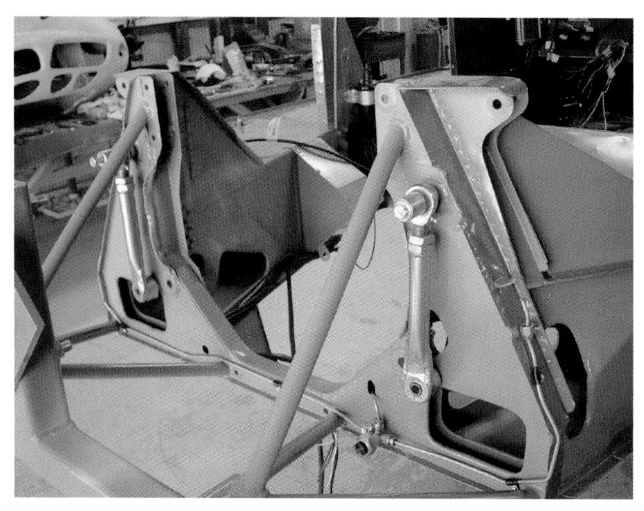

The rear bulkhead, suspension and brake lines being fitted. (Courtesy Marc Willard)

One Ford 289 V8, ready for fitting. (Courtesy Marc Willard)

The completed chassis with the engine, radiator, windshield, roof, front suspension and steering rack fitted. (Courtesy Marc Willard)

Getting there! (Courtesy Marc Willard)

The engine and transmission, now ready for installation. (Courtesy Marc Willard)

The pressure plate of the clutch assembly. (Courtesy Marc Willard)

The bellhousing and differential carrier of the Colotti transaxle. (Courtesy Marc Willard)

The Colotti gearbox, which proved to be the weak link on both the Lola Mk 6 GT and the Ford GT40. (Courtesy Marc Willard)

Two views of the straight-cut gears in the Colotti gearbox. (Courtesy Marc Willard)

The Colotti gearbox, now restored, plus other parts, such as the exhaust manifolds and driveshafts, and a rubber 'doughnut' coupling. (Courtesy Marc Willard)

Gearbox parts. (Courtesy Marc Willard)

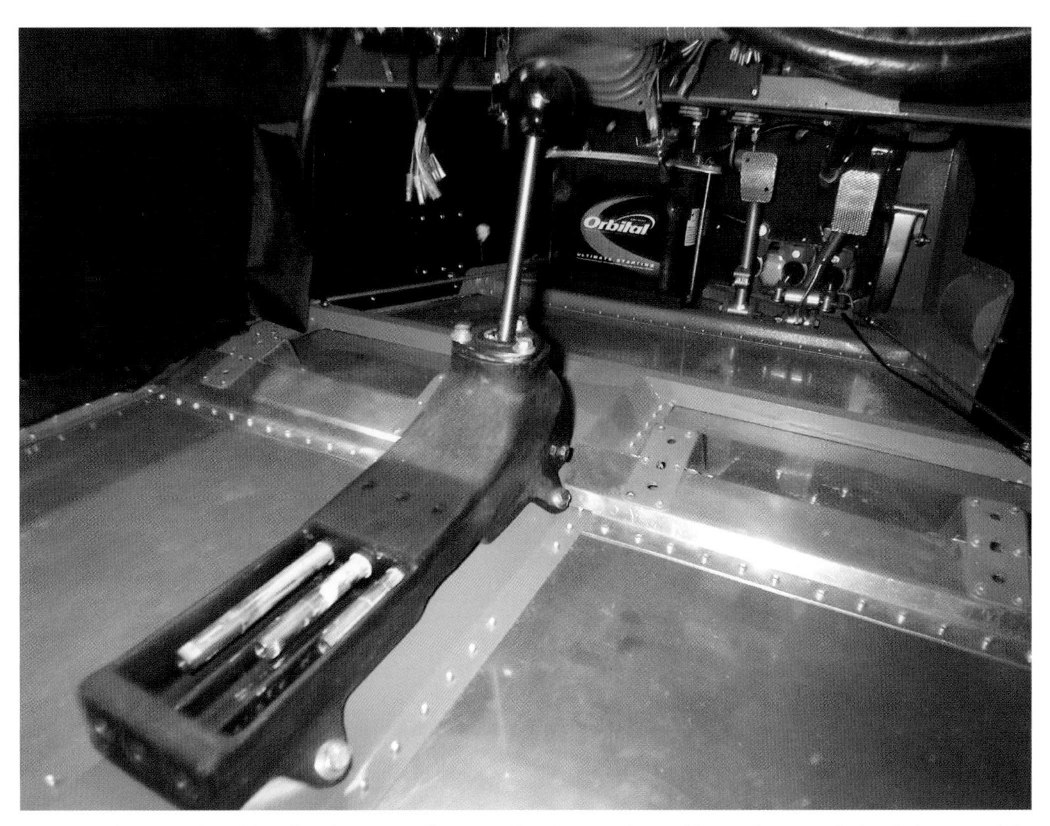

The gear lever and pedals. The three attachments for the Bowden cable can be seen behind the gearshift. (Courtesy Marc Willard)

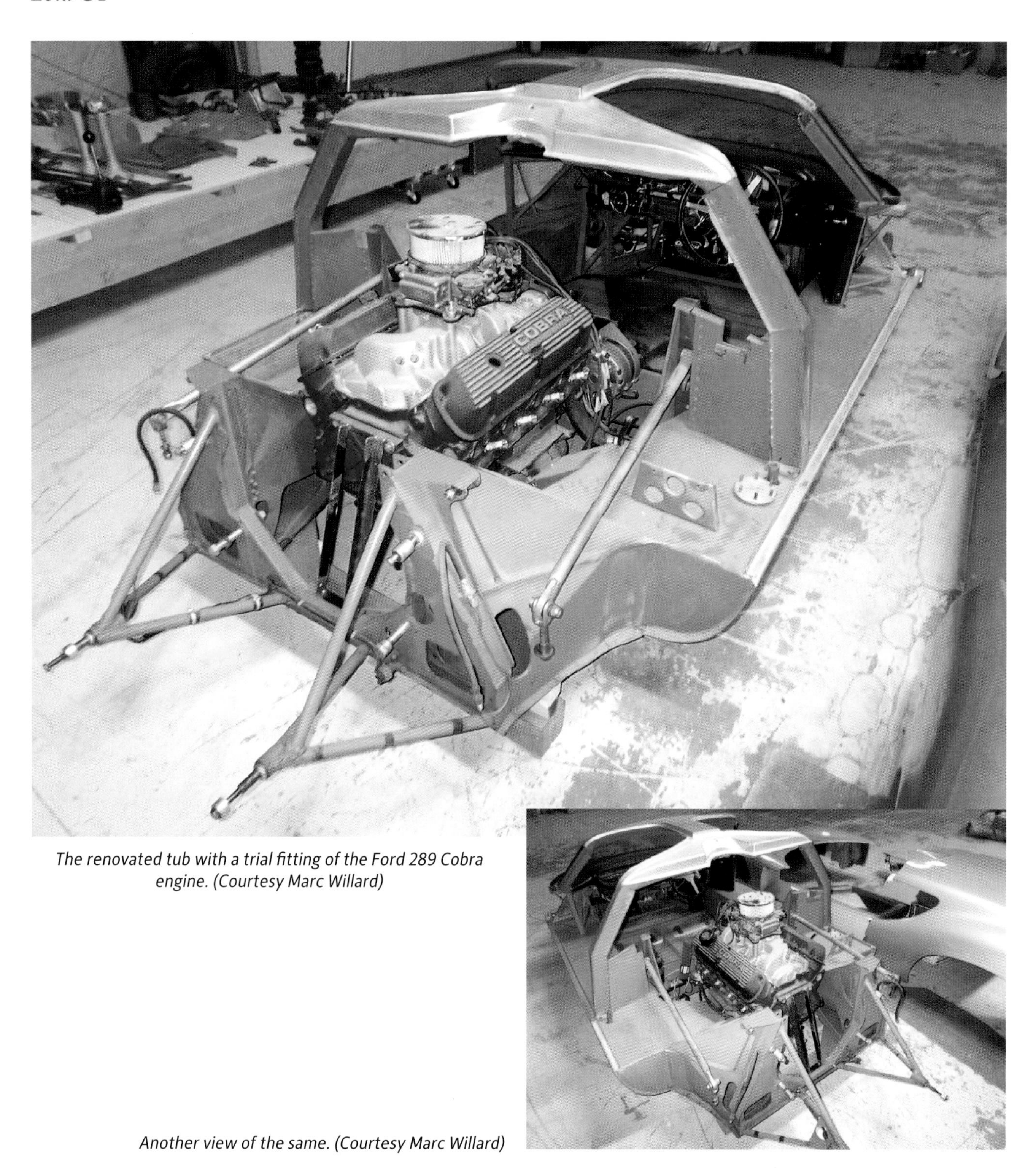

The renovated tub with a trial fitting of the Ford 289 Cobra engine. (Courtesy Marc Willard)

Another view of the same. (Courtesy Marc Willard)

Another view of the tub from above. (Courtesy Marc Willard)

The front compartment showing the steering rack and the fluid reservoirs. (Courtesy Marc Willard)

The right-hand side of the front of the chassis, showing the radiator and fluid reservoirs fitted. (Courtesy Marc Willard)

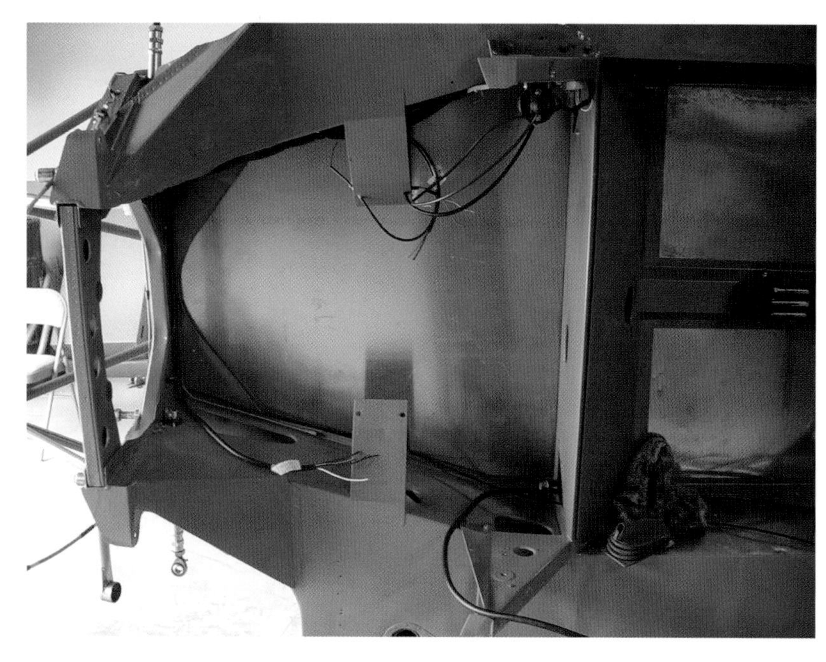

The front of the renovated chassis, showing parts beginning to be fitted. (Courtesy Marc Willard)

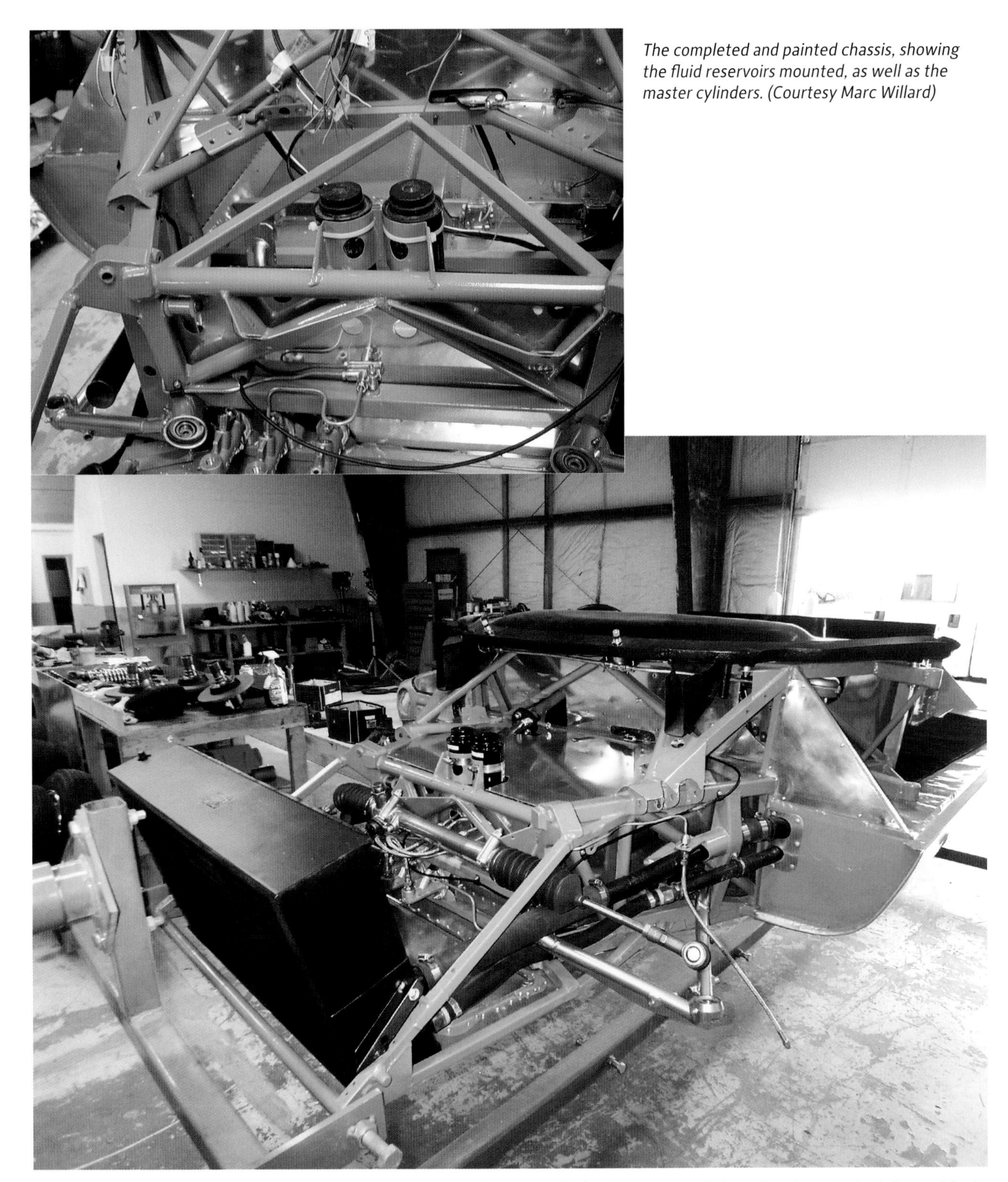

The completed and painted chassis, showing the fluid reservoirs mounted, as well as the master cylinders. (Courtesy Marc Willard)

The left-hand front of the renovated chassis, showing the radiator installed, with its water piping going through the left-hand fuel tank. (Courtesy Marc Willard)

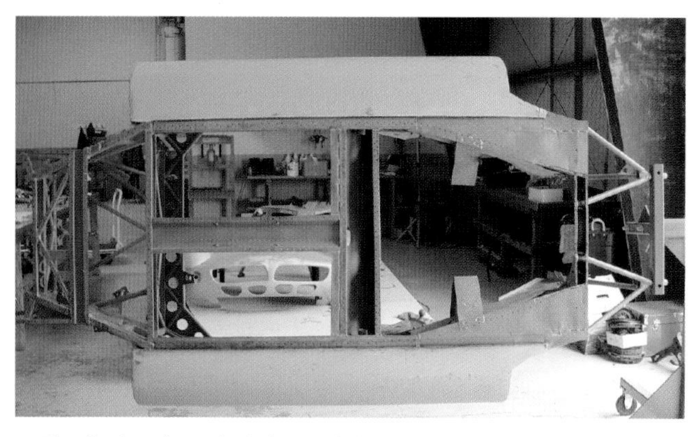

The fuel tanks, which formed the rocker panels/sills, are now mounted on the chassis, still in primer. (Courtesy Marc Willard)

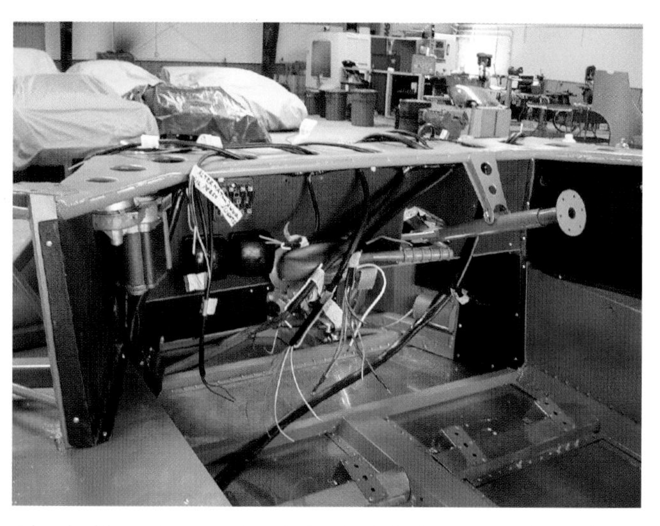

The dashboard area before being installed, showing the restored windshield wiper motor on the left-hand side. (Courtesy Marc Willard)

The newly installed dashboard mounting and gearshift installed. (Courtesy Marc Willard)

The dashboard area, showing the steering wheel installed and also the aluminium floor panels. (Courtesy Marc Willard)

Coil spring, Koni adjustable damper and lower wishbone fitted. The water piping to and from the radiator can also be seen. (Courtesy Marc Willard)

The rear of the restored chassis, showing the typical Lola rear bulkhead. (Courtesy Marc Willard)

Underneath the engine/rear cover of the Lola whilst under renovation. (Courtesy Marc Willard)

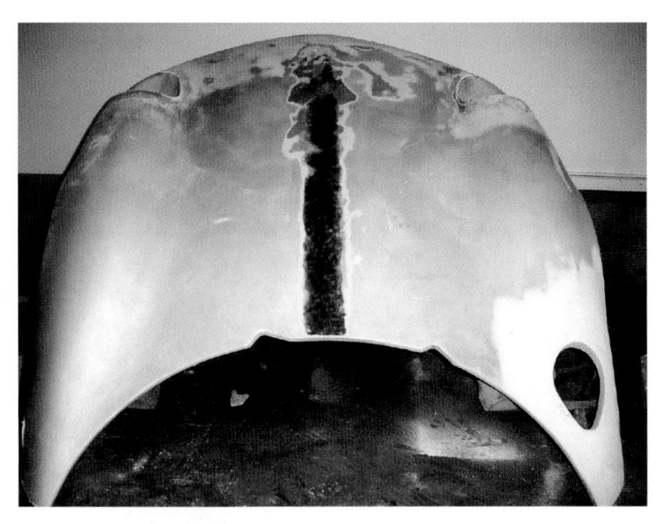

Two photos of the nose whilst under renovation. (Courtesy Marc Willard)

Almost ready! Body panels and chassis engine and gearbox, all ready for final assembly. (Courtesy Marc Willard)

Nearly there!: The restored and painted chassis on the left; the finished body panels on the right. (Courtesy Marc Willard)

The beautifully restored tail of the Lola Mk 6 GT. (Courtesy Marc Willard)

The body panels in a test fitting on the chassis. (Courtesy Marc Willard)

Another view of the same trial fitting. Note the cutaway in the roof for the doors. (Courtesy Marc Willard)

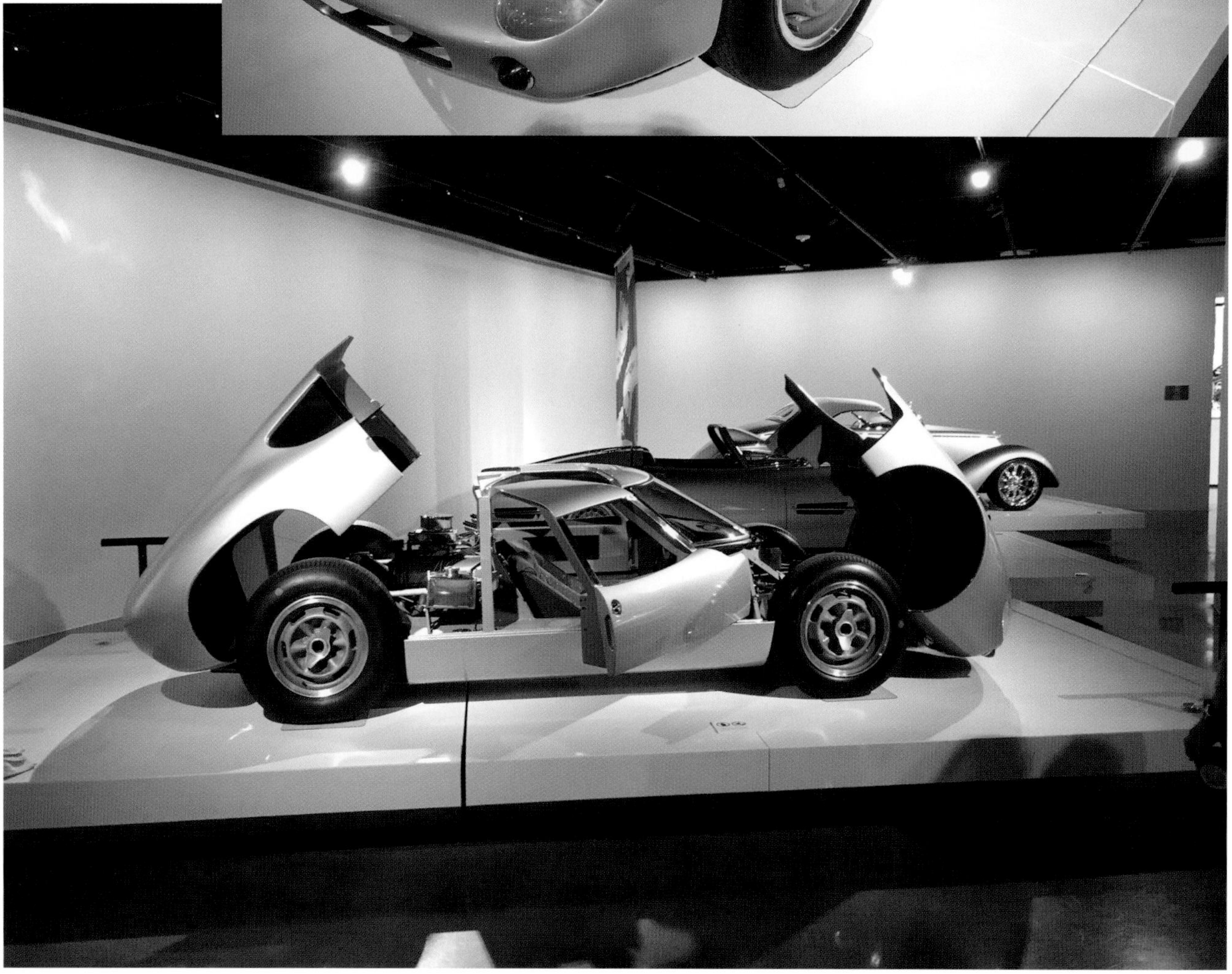

This page & opposite: The finished Lola GT Mk 6, LGT/P. (Courtesy Marc Willard)

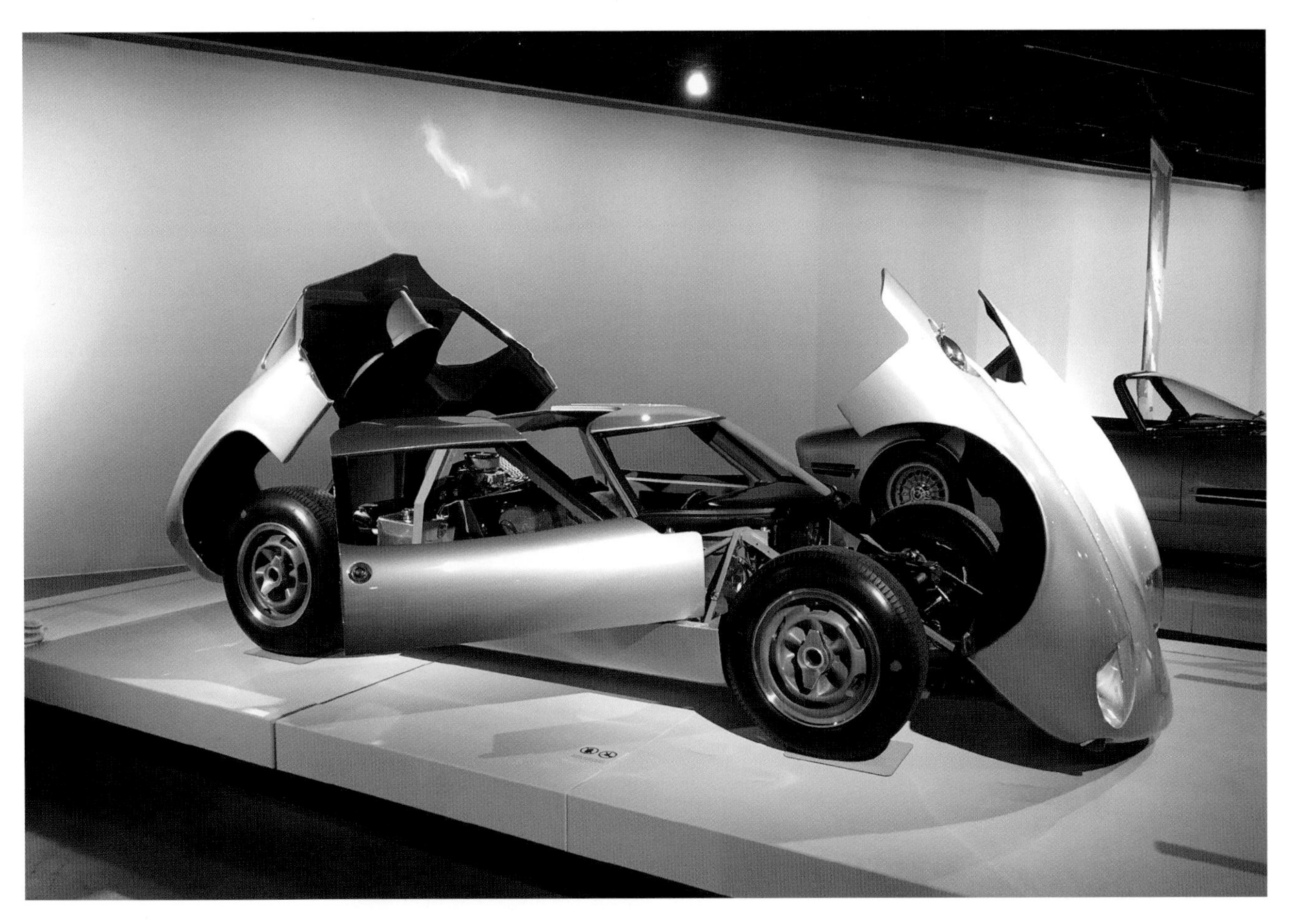

The starter motor of the Ford 289 engine.
(Courtesy Marc Willard)

The wheels of the Lola Mk 6 GT, beautifully restored and finished. (Courtesy Marc Willard)

A beautifully formed and restored A-arm/ wishbone; so typical of the exemplary restoration of this Lola. (Courtesy Marc Willard)

Rob Senekal (left), and Allen Grant. (Courtesy Marc Willard)

John Hill, the third partner in the restoration of the Lola Mk 6. (Courtesy Marc Willard)

Also from Veloce –

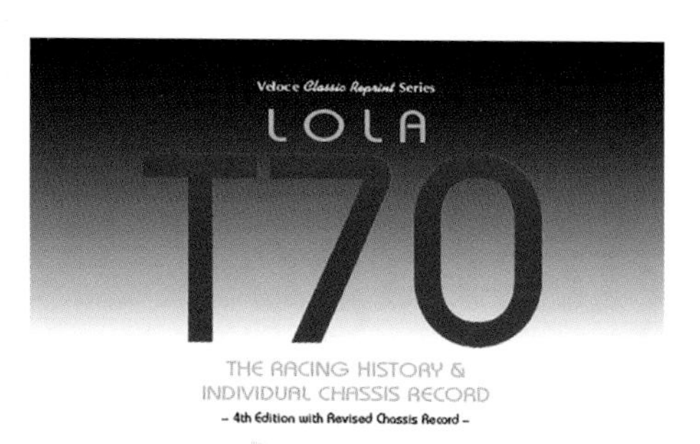

Reprinted after a long absence! The definitive development and racing history of the Lola T70. John Starkey has compiled a huge amount of information about the cars, and interviewed many past and present owners and drivers about their experiences with the T70. Contains the history and specification – where known – of each individual T70 chassis.

ISBN: 978-1-787110-51-9

Paperback • 25x20.7cm • 192 pages • 220 pictures

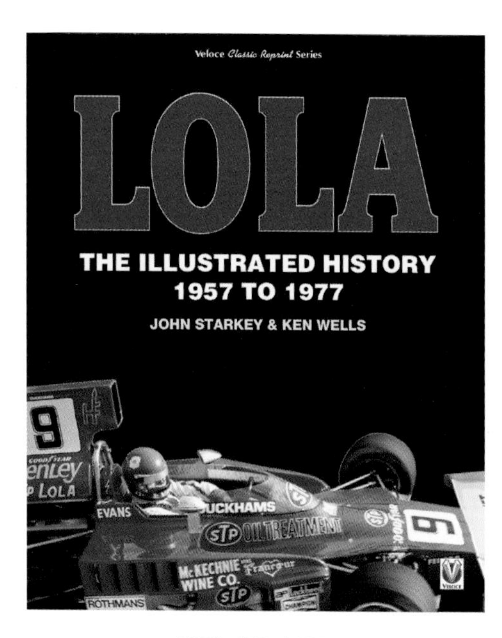

ISBN: 978-1-787111-04-2

Paperback • 25x20.7cm • 192 pages • 176 pictures

ISBN: 978-1-787112-58-2

Paperback • 25x20.7cm • 176 pages • 120 pictures

For more information and price details, visit our website at www.veloce.co.uk • email: info@veloce.co.uk • Tel: +44(0)1305 260068

ISBN: 978-1-787114-92-0

Paperback • 22.8x20.8cm • 192 pages • 162 pictures

ISBN: 978-1-787114-93-7

Paperback • 22.8x20.8cm • 128 pages • 74 pictures

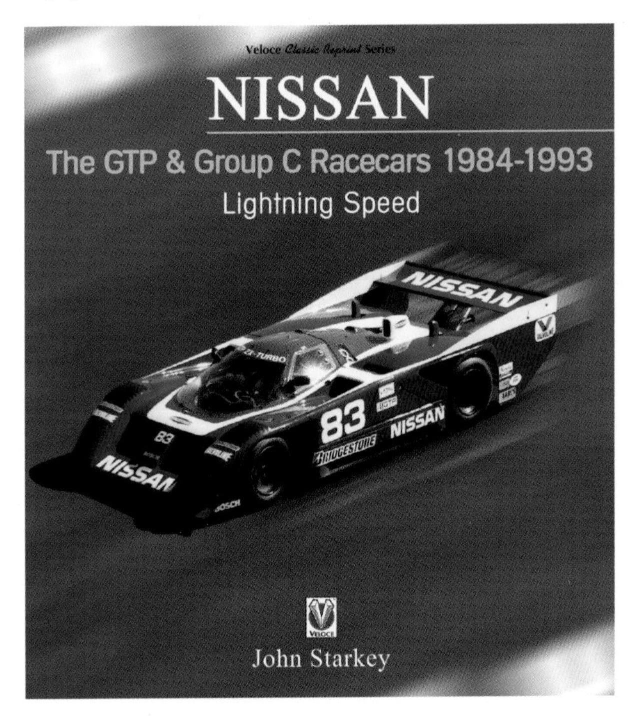

ISBN: 978-1-787114-94-4

Paperback • 22.8x20.8cm • 160 pages • 115 pictures

Also from Veloce –

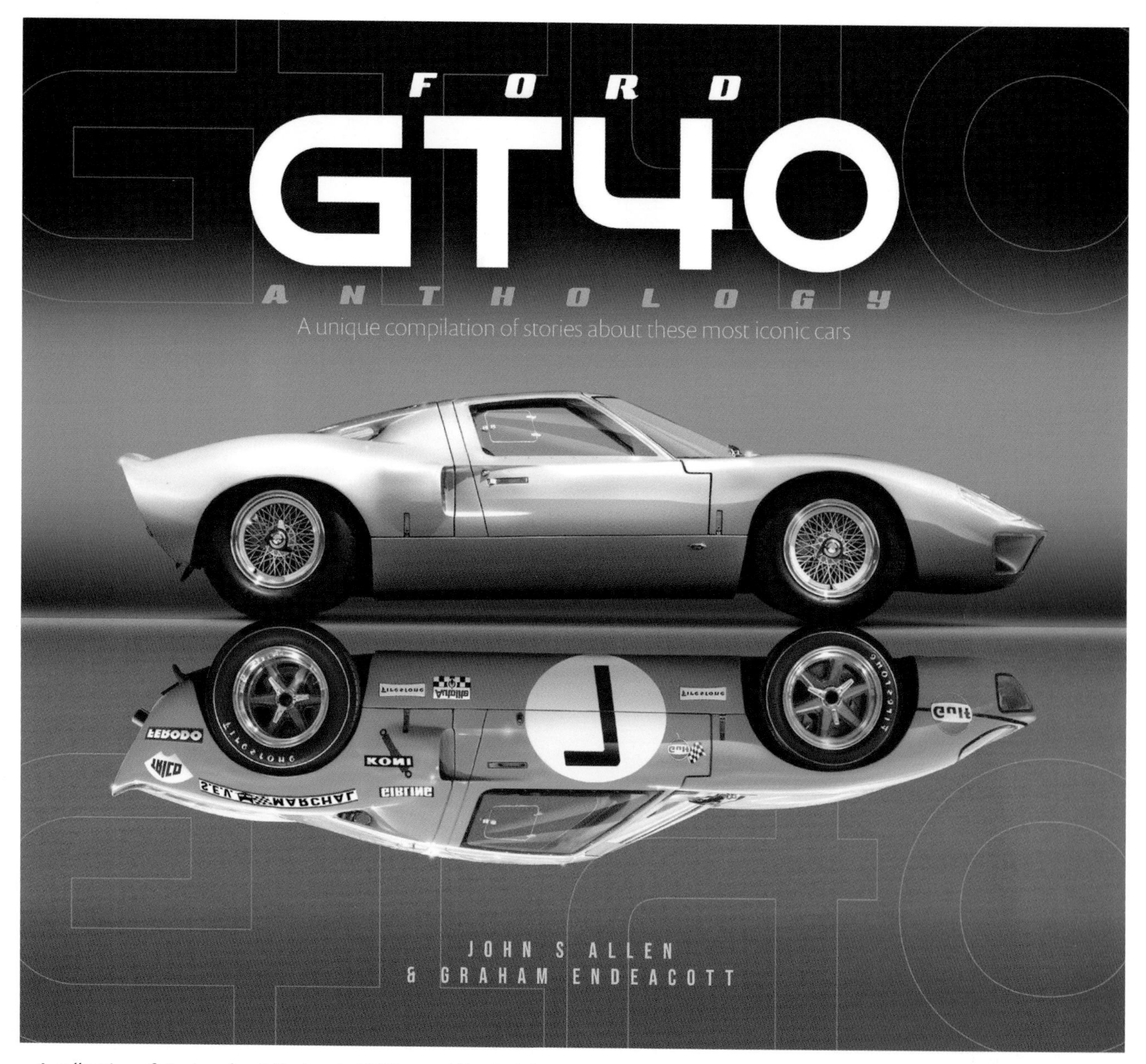

A collection of stories about the iconic GT40, providing insights to its design and racing achievements, and some well-kept secrets about its development. This is not a traditional history of the GT40, but a book of stand-alone stories that you can dip into, and also enjoy the many photographs that accompany the fascinating text.

ISBN: 978-1-787115-76-7

Hardback • 26.5x29.5cm • 320 pages • 500 pictures

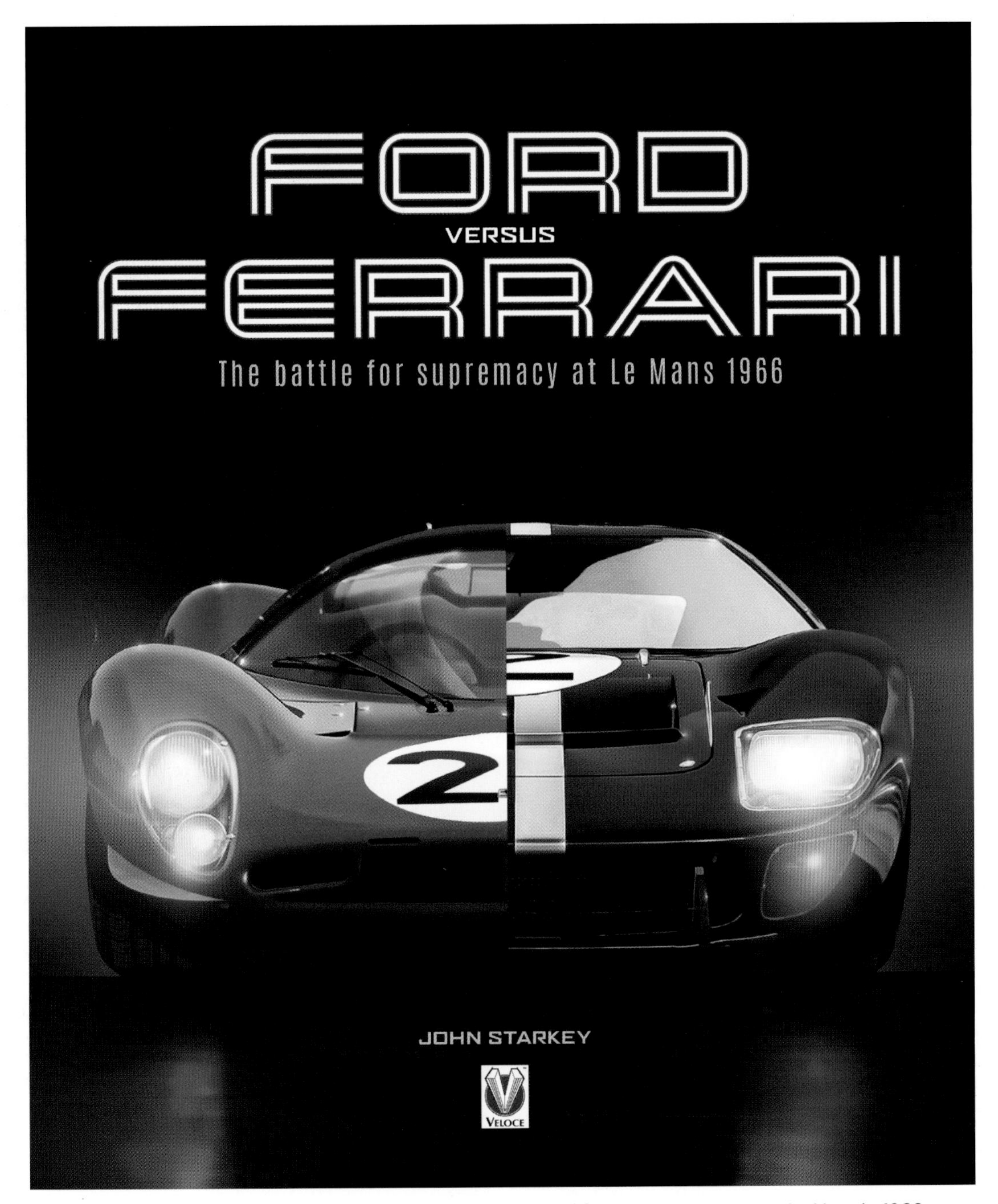
FORD
VERSUS
FERRARI
The battle for supremacy at Le Mans 1966
JOHN STARKEY
VELOCE

Index

AUTOSPORT
COPYRIGHT